Origination

RGS-IBG Book Series

For further information about the series and a full list of published and forthcoming titles please visit www.rgsbookseries.com

Published

Origination: The Geographies of Brands and Branding
Andy Pike

Frontier Regions of Marketization: Agribusiness, Farmers, and the Precarious Making of Global Connections in West Africa
Stefan Ouma

Africa's Information Revolution: Technical Regimes and Production Networks in South Africa and Tanzania
James T. Murphy and Pádraig Carmody

In the Nature of Landscape: Cultural Geography on the Norfolk Broads
David Matless

Geopolitics and Expertise: Knowledge and Authority in European Diplomacy
Merje Kuus

Everyday Moral Economies: Food, Politics and Scale in Cuba
Marisa Wilson

Material Politics: Disputes Along the Pipeline
Andrew Barry

Fashioning Globalisation: New Zealand Design, Working Women and the Cultural Economy
Maureen Molloy and Wendy Larner

Working Lives – Gender, Migration and Employment in Britain, 1945–2007
Linda McDowell

Dunes: Dynamics, Morphology and Geological History
Andrew Warren

Spatial Politics: Essays for Doreen Massey
Edited by David Featherstone and Joe Painter

The Improvised State: Sovereignty, Performance and Agency in Dayton Bosnia
Alex Jeffrey

Learning the City: Knowledge and Translocal Assemblage
Colin McFarlane

Globalizing Responsibility: The Political Rationalities of Ethical Consumption
Clive Barnett, Paul Cloke, Nick Clarke & Alice Malpass

Domesticating Neo-Liberalism: Spaces of Economic Practice and Social Reproduction in Post-Socialist Cities
Alison Stenning, Adrian Smith, Alena Rochovská and Dariusz Świątek

Swept Up Lives? Re-envisioning the Homeless City
Paul Cloke, Jon May and Sarah Johnsen

Aerial Life: Spaces, Mobilities, Affects
Peter Adey

Millionaire Migrants: Trans-Pacific Life Lines
David Ley

State, Science and the Skies: Governmentalities of the British Atmosphere
Mark Whitehead

Complex Locations: Women's Geographical Work in the UK 1850–1970
Avril Maddrell

Value Chain Struggles: Institutions and Governance in the Plantation Districts of South India
Jeff Neilson and Bill Pritchard

Queer Visibilities: Space, Identity and Interaction in Cape Town
Andrew Tucker

Arsenic Pollution: A Global Synthesis
Peter Ravenscroft, Hugh Brammer and Keith Richards

Resistance, Space and Political Identities: The Making of Counter-Global Networks
David Featherstone

Mental Health and Social Space: Towards Inclusionary Geographies?
Hester Parr

Climate and Society in Colonial Mexico: A Study in Vulnerability
Georgina H. Endfield

Geochemical Sediments and Landscapes
Edited by David J. Nash and Sue J. McLaren

Driving Spaces: A Cultural-Historical Geography of England's M1 Motorway
Peter Merriman

Badlands of the Republic: Space, Politics and Urban Policy
Mustafa Dikeç

Geomorphology of Upland Peat: Erosion, Form and Landscape Change
Martin Evans and Jeff Warburton

Spaces of Colonialism: Delhi's Urban Governmentalities
Stephen Legg

People/States/Territories
Rhys Jones

Publics and the City
Kurt Iveson

After the Three Italies: Wealth, Inequality and Industrial Change
Mick Dunford and Lidia Greco

Putting Workfare in Place
Peter Sunley, Ron Martin and Corinne Nativel

Domicile and Diaspora
Alison Blunt

Geographies and Moralities
Edited by Roger Lee and David M. Smith

Military Geographies
Rachel Woodward

A New Deal for Transport?
Edited by Iain Docherty and Jon Shaw

Geographies of British Modernity
Edited by David Gilbert, David Matless and Brian Short

Lost Geographies of Power
John Allen

Globalizing South China
Carolyn L. Cartier

Geomorphological Processes and Landscape Change: Britain in the Last 1000 Years
Edited by David L. Higgitt and E. Mark Lee

Forthcoming

Smoking Geographies: Space, Place and Tobacco
Ross Barnett, Graham Moon, Jamie Pearce, Lee Thompson and Liz Twigg

Home SOS: Gender, Injustice and Rights in Cambodia
Katherine Brickell

Nothing Personal? Geographies of Governing and Activism in the British Asylum System
Nick Gill

Pathological Lives: Disease, Space and Biopolitics
Steve Hinchliffe, Nick Bingham, John Allen and Simon Carter

Work–Life Advantage: Sustaining Regional Learning and Innovation
Al James

Rehearsing the State: The Political Practices of the Tibetan Government-in-Exile
Fiona McConnell

Articulations of Capital: Global Production Networks and Regional Transformations
John Pickles, Adrian Smith and Robert Begg, with Milan Buček, Rudolf Pástor and Poli Roukova

Body, Space and Affect
Steve Pile

Making Other Worlds: Agency and Interaction in Environmental Change
John Wainwright

Everyday Peace? Politics, Citizenship and Muslim Lives in India
Philippa Williams

Metropolitan Preoccupations: The Spatial Politics of Squatting in Berlin
Alexander Vasudevan

Origination

The Geographies of Brands and Branding

Andy Pike

WILEY Blackwell

This edition first published 2015
© 2015 John Wiley & Sons, Ltd.

Registered Office
John Wiley & Sons, Ltd, The Atrium, Southern Gate, Chichester, West Sussex, PO19 8SQ, UK

Editorial Offices
350 Main Street, Malden, MA 02148-5020, USA
9600 Garsington Road, Oxford, OX4 2DQ, UK
The Atrium, Southern Gate, Chichester, West Sussex, PO19 8SQ, UK

For details of our global editorial offices, for customer services, and for information about how to apply for permission to reuse the copyright material in this book please see our website at www.wiley.com/wiley-blackwell.

The right of Andy Pike to be identified as the author of this work has been asserted in accordance with the UK Copyright, Designs and Patents Act 1988.

All rights reserved. No part of this publication may be reproduced, stored in a retrieval system, or transmitted, in any form or by any means, electronic, mechanical, photocopying, recording or otherwise, except as permitted by the UK Copyright, Designs and Patents Act 1988, without the prior permission of the publisher.

Wiley also publishes its books in a variety of electronic formats. Some content that appears in print may not be available in electronic books.

Designations used by companies to distinguish their products are often claimed as trademarks. All brand names and product names used in this book are trade names, service marks, trademarks or registered trademarks of their respective owners. The publisher is not associated with any product or vendor mentioned in this book.

Limit of Liability/Disclaimer of Warranty: While the publisher and author have used their best efforts in preparing this book, they make no representations or warranties with respect to the accuracy or completeness of the contents of this book and specifically disclaim any implied warranties of merchantability or fitness for a particular purpose. It is sold on the understanding that the publisher is not engaged in rendering professional services and neither the publisher nor the author shall be liable for damages arising herefrom. If professional advice or other expert assistance is required, the services of a competent professional should be sought.

Library of Congress Cataloging-in-Publication Data applied for.

C: 9781118556382
P: 9781118556405

A catalogue record for this book is available from the British Library.

Cover image: © Michelle Wood

Set in 9.5/11.5pt Plantin by SPi Publisher Services, Pondicherry, India
Printed and bound in Malaysia by Vivar Printing Sdn Bhd

1 2015

For my parents, Michelle, Ella and Connell

Contents

Series Editors' Preface

The RGS-IBG Book Series only publishes work of the highest international standing. Its emphasis is on distinctive new developments in human and physical geography, although it is also open to contributions from cognate disciplines whose interests overlap with those of geographers. The Series places strong emphasis on theoretically informed and empirically strong texts. Reflecting the vibrant and diverse theoretical and empirical agendas that characterize the contemporary discipline, contributions are expected to inform, challenge and stimulate the reader. Overall, the RGS-IBG Book Series seeks to promote scholarly publications that leave an intellectual mark and change the way readers think about particular issues, methods or theories.

For details on how to submit a proposal please visit:
www.rgsbookseries.com

Neil Coe
National University of Singapore

Tim Allott
University of Manchester, UK

RGS-IBG Book Series Editors

Acknowledgements

While books are individually authored, they are collective efforts. Thanks are due to many people who have helped the production of this book: Neil Coe, Series Editor, for his advice, encouragement and perceptive feedback on the first draft; the book proposal reviewers; the book draft manuscript reviewers; Jacqueline Scott at Wiley-Blackwell; the participants in the research undertaken for the book; Charlie Thompson for the research assistance; the British Academy for funding the research on Burberry and the research assistance of Angela Abbott, Pedro Marques and Jon Swords; the Economic and Social Research Council (ESRC) for funding and the NewcastleGateshead Initiative and, specifically, Tina Snowball for supporting the ongoing research on space and place branding and reputation and the work of Rebecca Richardson and Fraser Bell; Michelle Wood for the cover art; the editors and reviewers for journals where the formative ideas have been published, especially Roger Lee and David Rigby; Stuart Dawley, Danny MacKinnon, Phil O'Neill and John Tomaney for their comments on draft chapters; the participants and contributors to the seminars and conference sessions in Aalborg, Beijing, Boston, Edinburgh, Glasgow, Lisbon, London, Manchester, Newcastle, Nottingham and Sheffield where the ideas have been introduced, explored and refined – especially John Allen, Nicola Bellini, Andy Cumbers, Stuart Dawley, Andy Gillespie, Henrik Halkier, Ray Hudson, Alex Hughes, Guy Julier, Damian Maye, Kevin Morgan, Liz Moor, Cecilia Pasquinelli, Jane Pollard, Dominic Power, Ranald Richardson, Andrés Rodríguez-Pose and Henry Yeung; and colleagues at the Centre for Urban and Regional Development Studies (CURDS), Newcastle University, UK, which continues to provide a distinctive and stimulating research culture and outlook that has inspired and inflected this book. The insights and questions of the PhD and MA postgraduates in the local and regional development programmes in CURDS and undergraduates on the geography programmes at Newcastle University have further contributed to sharpening the understanding and communication of origination in brand and branding geographies. The usual disclaimers apply.

Permissions

We are grateful to those listed for permission to reproduce copyright material:

Sarah Mulligan, Newcastle City Libraries, for Figure 1.1;
Catherine Scott-Dunkes, Baltimore Museum of Industry, for Figure 1.3;
Vincenzo Mammola, Prada, for Figure 1.4 and Figure 3.6;
Steve Lovelace for Figure 1.8;
David Wengrow, UCL, for Figure 2.5;
Sebastian Miroudot, World Bank, for Figure 3.2;
The BBC for Figure 3.6;
Scottish and Newcastle for Figures 4.3, 4.6 and 4.7;
Jygsaw Brands for Figure 4.5;
Sharon McKee, NewcastleGateshead Initiative, for Figure 4.6;
Leighton Andrews, AM Welsh Assembly Government, for Figure 5.4;
Kim McDonald, Into Somerset, for Figure 7.1;
Jackson Tucker Lynch and the Harris Tweed Authority for Figure 7.4;
Christian Scully and Corbis for Figure 7.7.

An effort has been made to contact copyright holders for their permission to reprint material in this book. The publisher would be grateful to hear from any copyright holder who is not here acknowledged and will undertake to rectify any errors or omissions in future editions of this book.

List of Tables

List of Figures

Chapter One
Introduction

Introduction: Where are goods and services commodities from and why does it matter?

From the regional heyday of producing a quarter of the world's ships in the opening decade of the twentieth century (Hudson 1989), Tyneside in north east England established a reputation for engineering innovation and manufacturing prowess. The 'carboniferous capitalism' of coal, iron and steel underpinned specialization and international technological leadership in heavy engineering in Britain's imperial markets (Tomaney 2006). Industrial pioneers such as William Armstrong, Charles Parsons and George Stephenson in concert with skilled and unionized urban labour meant 'Made in Tyneside' was commercially meaningful and valuable (Middlebrook 1968). During the 1950s and 1960s, Historian Paul Kennedy described this time and place as:

> A world of great noise and much dirt... [where] ... There was a deep satisfaction about making things ... among all of those that had supplied the services, whether it was the local bankers with credit; whether it was the local design firms. When a ship was launched at Swan Hunter [Wallsend, North Tyneside] all the kids at the local school went to see the thing our fathers had put together and when we looked down from the cross-wired fence, tried to find Uncle Mick, Uncle Jim or your dad, this notion of an integrated, productive community was quite astonishing (quoted in Chakrabortty 2011: 1).

Origination: The Geographies of Brands and Branding, First Edition. Andy Pike.
© 2015 John Wiley & Sons, Ltd. Published 2015 by John Wiley & Sons, Ltd.

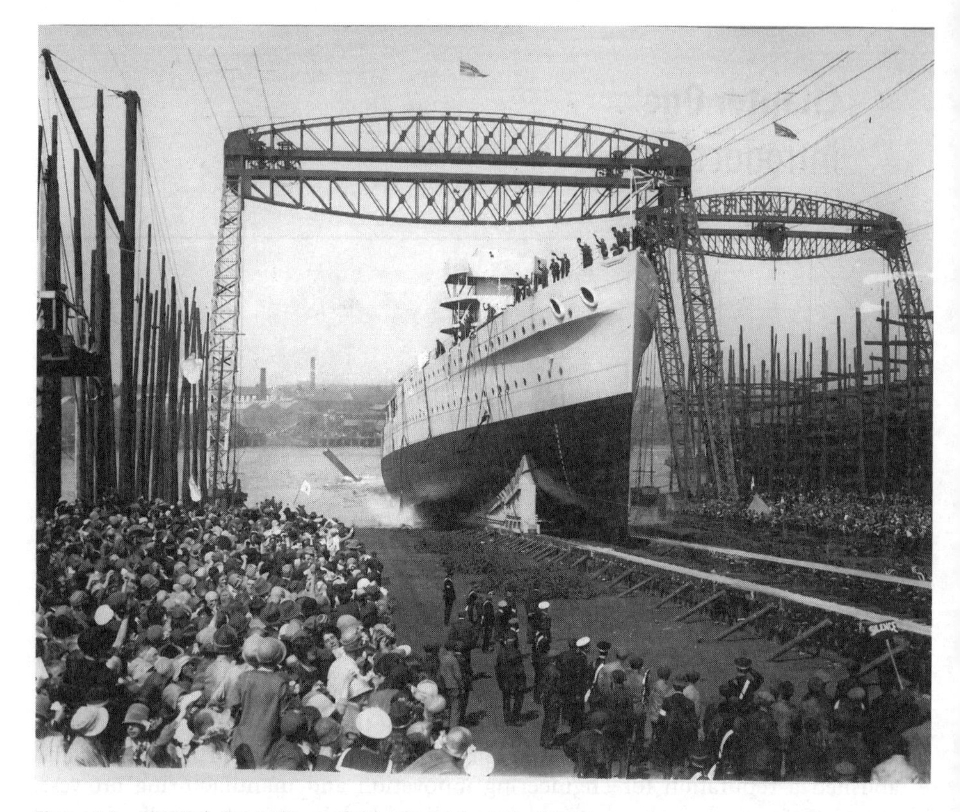

Figure 1.1 HMS York. Source: Newcastle Libraries & Information Service.

Vessels, such as HMS York (Figure 1.1), were made in the shipyards of Hebburn, Walker and Wallsend, and, once departed from the slipway, travelled the world as functional commodities embodying the meaning and commercially valuable reputation of where they were from and who built them.

Although Tyneside has since been ravaged by waves of deindustrialization and a highly socially and spatially uneven transition to a service-dominated economy (Pike *et al.* 2006), the geographical associations in what a place is known for live on in certain specialist market niches. In the kinds of connections, for example, made in the corporate logo of Tyneside Safety Glass, including a silhouette of the Tyne Bridge, and the marking of some of its products with the slogan 'Tyneside Toughened'. Tyneside Safety Glass is a privately owned specialist glass processor established in 1937 with its headquarters in the Team Valley south of the river Tyne in Gateshead. It employs around 200 people and operates three factories in north east England. The company articulates authentic claims to provenance as part of its creation and communication of meaning and value for its customers in international architectural, automotive, defence and security markets. There are no intrinsic ties that mean such goods and services commodities could not technically be produced elsewhere beyond Tyneside in north east England. But commercial advantage is being sought by the owners through the company name, logo and slogan making strong and geographical connections to the historical traditions, character and reputation of the place of Tyneside for engineering ingenuity, technological innovation and manufacturing precision.

As Tyneside Safety Glass demonstrates, where goods and services commodities are from and are associated with – and, crucially, are *perceived* to be from and associated with – and why is important. Raising such issues encourages reflection upon how we understand and explain critical spatial concerns about the geographies of economy and their organization and dynamics: the call centres, design studios, factories, laboratories, logistics hubs, market stalls, offices, shops, trading floors, warehouses and the investments, jobs, incomes, livelihoods and identities in cities, localities, regions and countries with which they are entwined. Such concerns make us think about how, why, where and by whom goods and services commodities are associated with specific and particular geographical attributes and characteristics of spaces and places as part of attempts by myriad actors to create meaning and value.

Longstanding connections and connotations are evident especially where the geographical associations of goods and services commodities are strong, enduring and decisive commercial and trade advantages. Well known examples include 'Danish furniture, Florentine leather goods, Parisian *haute couture*, Champagne wines, London theatre, Swiss watches before digitization, Thai silks, recorded music from Nashville ... Hollywood films' (Scott 1998: 109). The list could go on. For over four decades, researchers in the discipline of marketing have recognized this phenomenon and call it the '*Country* of Origin' effect (Bass and Wilkie 1973). By this, they mean the consumer views of the different capabilities and historical reputations of countries for particular goods and services. These perceptions influence consumer assessments of attributes such as quality, style and taste, and interpretation of meaning and value that shapes their purchasing decisions (Phau and Prendergast 2000). Importantly, these geographical associations and reputations tend to be sticky, slow changing and, once accumulated, can become difficult to change or dislodge. As Harvey Molotch (2002: 677) puts it, 'perfume should come from Paris not Peoria, watches from Geneva not Gdansk'. Such geographical associations are powerful in the ways in which they create – and potentially destroy – meaning and value through what they explicitly demonstrate or imply for specific goods and services commodities in particular spatial and temporal market contexts.

The origins of brands and branding

Historically, goods and later services commodities bore marks or brands as means of distinction from competitors and signs of quality and reliability (Room 1998). Artisanal producers in ancient Greece and Rome marked their goods such as pottery with distinctive signs to communicate their origin and quality (Lindemann 2010). Individual marks or seals that identified particular craft producers or traders were evident c.300 BC. Merchants initially used generic symbols to communicate the business in which they traded, including 'a ham for butchers, a cow for creameries' (Chevalier and Mazzolovo 2004: 15). Makers' marks began evolving into brands and became more evident and important during the seventeenth and eighteenth centuries. This development involved especially craft goods such as furniture, porcelain and tapestries, particularly when travelling for sale beyond face-to-face transactions in localized markets (Room 1998). As David Wengrow (2008: 21) argues, 'commodity branding':

> has been a long-term feature of human cultural development, acting within multiple ideological and institutional contexts including those of sacred hierarchies and sacrificial economies of a certain scale. What *has* varied significantly over time and space is the nexus of authenticity, quality control, and desire from which brand economies draw their authority; the web of agencies (real or imagined) through which homogeneous goods must be seen to pass in order to be consumed, be they the bodies of the ancestral dead, the gods, heads of state, secular business gurus, media celebrities, or that core fetish of post-modernity, the body of the sovereign consumer citizen in the act of self-fashioning (emphasis in original).

Industrialization and mass production in the nineteenth century underpinned and reinforced the commercial value and meaning of branding, especially for packaged goods: 'Through industrialization the production of many household items, such as soap, moved from local production to centralized factories. As the distance between buyer and supplier widened the communication of origin and quality became more important' (Lindemann 2010: 3). The naming of 'Platt's *Brand* Raw Oysters' and the explicit use of the term brand in the advertising of 'Jackson Square Cigar – America's Standard 5¢ *Brand*' as particular kinds of commodities demonstrate the early and explicit incorporation of the term 'brand' into product names and their circulation and promotion (Figure 1.2). Mass production and distribution generated economies of scale and lowered production costs, but required mass markets and the communication and demonstration of superior quality to dislodge local consumer preferences for local producers.

The etymological roots of the word brand as a noun lie in several linguistic traditions. These refer commonly to a fire or flame as well as firebrand, piece of burning wood and torch: the Old English of *brand* and *brond*; the Old Norse *brandr*; the Old High German *brant*; the Old Frisian *brond* and the German *brand* (Collins Concise Dictionary Plus 1989). Historically, from around the 1550s, as a noun a brand was defined as an identifying mark to signify ownership burned on livestock as well as criminals and slaves with a branding iron. With the emergence of craft production and later industrialization, brand became defined as a type or kind of good or service from a specific company sold under a particular name, often referred to as its 'brand name' and encapsulating a particular design, identity and/or image. As a verb, from the 1400s, to brand meant to mark, to cauterize – often wounds – and to stigmatize typically criminals and slaves. From the 1580s, the meaning of the verb evolved to refer to the marking of property and ownership.

Figure 1.2 'Platt's Brand Raw Oysters' and 'Jackson Square Cigar – America's Standard 5¢ Brand'. Source: Historical images from Baltimore Museum of Industry.

Branding emerged as a process that tries to articulate, integrate and enhance the attributes embodied and connected in brands in meaningful and valuable ways. Jan Lindemann (2010: 3) describes how:

> Although the initial purpose of branding was to demonstrate the origin of an animal it quickly grew into a means of differentiation. Over time a farmer would establish a certain reputation for the quality of his cattle expressed by the branded mark on the animal. This enabled buyers quickly to assess the quality of the cattle and the price they were willing to pay for it.

Branding developed rapidly to become part of connecting meaning and value through associations across a wider range of goods and services. Branding has underpinned the process of brand extension by actors into certain spatial and temporal market settings. Examples include Italian fashion house Prada's excursion into the mobile phone business with LG, and UK supermarket Tesco's development of Tesco Bank financial services (Figure 1.3). In the era of industrialization and mass production and consumption, branding sat within Raymond Williams' (1980: 184) broader definition of advertising as 'a highly organized and professional system of magical inducements and satisfactions, functionally very similar to magical systems in simpler societies, but rather strangely co-existent with a highly developed scientific technology'.

Figure 1.3 Brand extension: Prada and LG mobile phone and Tesco Bank financial services. Source: Prada SA; Tesco Bank.

The rise of brands and branding

In the transition from a producer to a consumer-dominated economy, society, culture, ecology and polity (Bauman 2007), the brands and branding of goods and services commodities have risen to prominence in dramatic fashion. Brands and branding have proliferated. In the United Kingdom alone, the number of brands has risen from an estimated 2 million in 1997 to over 8 million in 2011 in a marketing context in which '80% of categories are seen as increasingly homogenous'; amidst the proliferation of media channels in the digital era consumers are being 'bombarded with up to 5,000 marketing messages every day' (Noble 2011: 29). Brands were traditionally treated in accounting as 'goodwill': the difference between the purchase price of a business and the book value of its assets (Lindemann 2010). Brands have now increased sufficiently in importance to become explicitly recognized as economic entities necessitating calculation of their financial value and incorporation into corporate accounts. As Jan Lindemann (2010: 5) explains:

> In financial terms, the brand constitutes an intangible asset that provides its owners with an identifiable and ownable cash flow over the time of its useful economic life. This can span more than 100 years as evidenced by brands such Coca-Cola, Nokia, and Goldman Sachs. The brand is an economic asset that creates cash flows on a stand-alone basis (e.g. licensing) or integrated with other tangible and intangible assets. The mental impact of branding is only economically relevant if it results in a positive financial return

Table 1.1 Brand valuation methodologies, 2009 ($m)

Brand	*Business Week Interbrand*	*Milward Brown*	*Brand finance*	*Brand value average*	*% of market capital*
Coca-Cola	68 734	67 625	32 728	56 362	49
IBM	60 211	66 662	31 530	52 801	34
GE	47 777	59 793	26 654	44 741	30
Nokia	34 864	35 163	19 889	29 972	74
Apple	15 433	63 113	13 648	30 731	21
McDonald's	32 275	66 575	200 003	39 618	65
HSBC	10 510	19 079	25 364	18 318	17
American Express	14 971	14 963	9944	13 293	37
Google	31 980	100 039	29 261	53 760	38
Nike	13 179	11 999	14 583	13 254	48

Note: [a]Nominal prices.
Source: Adapted from Lindemann (2010: 10).

> for the user or owner of the brand that outstrips the investments into the brand. The impact of brands on shareholder value is substantial and can amount up to 80 per cent of shareholder value.

In acquisition, merger and takeover activities, the difference between the purchase price of the company and the value of its material or tangible assets has been attributed to the intangible asset of the brand (Lindemann 2010). Brands and branding have become critical sources of often enduring economic meaning and value, integral to shaping the agency of actors involved in corporate and industrial strategies internationally.

Amidst competition amongst consultancies offering proprietary methodologies, brands are now valued and ranked. Specific techniques such as Interbrand's 'Best Global Brand', Millward Brown's 'Brand Dynamics/BrandZ', Brand Finance's 'Brand Valuation' and Young and Rubicam's 'BrandAsset Evaluator' attribute different values to particular brands (Table 1.1) (Lindemann 2010). Derided as the 'professional persuaders' in Vance Packard's (1980: 31) classic book, media holding companies providing assorted advertising, branding and media planning services are now amongst the world's largest companies. Market leader the WPP group grossed over $14 billion in sales revenues in 2010 (Figure 1.4) (see Faulconbridge *et al.* 2011). As the media landscape has fragmented, splintered and proliferated across emergent technologies and multiple channels (e.g. billboard, on-line, print, radio, social media, TV), media planning companies working with brand owners and managers to place and position their brands have grown in importance, size and value (Kornberger 2010). While it is difficult accurately to count the complete volume and value of activity in the world of brands and branding, Liz Moor's (2008: 413) analysis concludes that 'branding is an increasingly significant component of the design

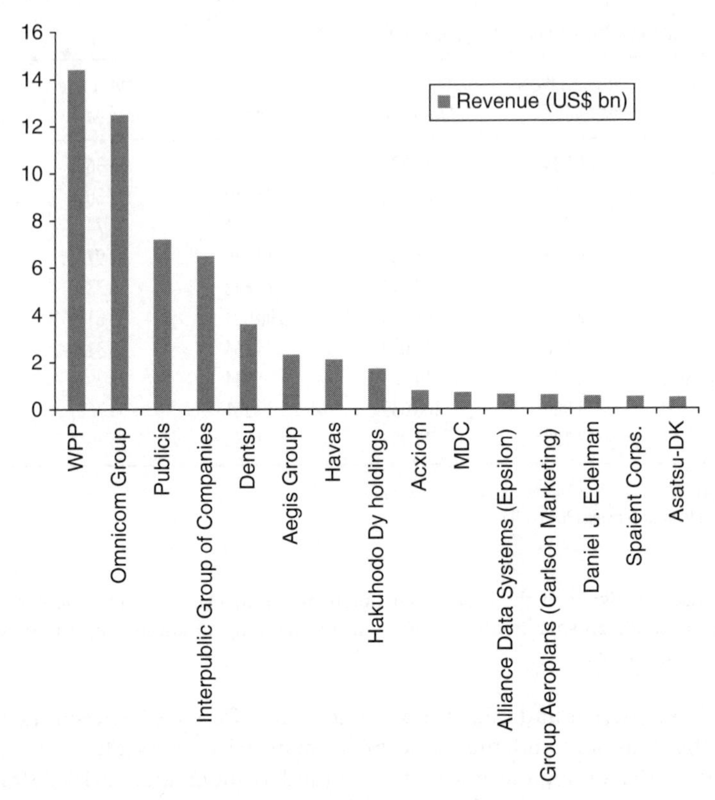

Figure 1.4 Global advertising agencies by revenue (US$bn), 2010. Source: Calculated from AdAge data. Note: [a]Nominal prices.

industry in Britain, while design itself is one of the largest sectors within the "creative industries"'. Liz Moor (2008: 415, emphasis in original) further notes how 'Part of what distinguishes branding from advertising is its extended spatial scope and broad conception of the potential *media* for commercial communication' such that 'corporate identity and branding consultancies had finally come close to realizing James Pilditch's original aspiration of becoming not simply an adjunct of advertising, but rather "the new total"'.

The dramatic rise, pervasiveness and importance of brands and branding in contemporary economy, society, culture, ecology and polity has been widely recognized. Martin Kornberger (2010: xi) interprets the emergence of a 'brand society' wherein brands as 'ready-made identities' are 'so mashed up with our social world that they have become a powerful life-shaping force'. He goes on to claim that brands may be 'the most ubiquitous and pervasive cultural form in our society' that are 'rapidly becoming one of the most powerful of the phenomena transforming the way we manage organizations and live our lives' (Kornberger 2010: xii, 23). Adam

Arvidsson (2005: 236) too interprets 'a well nigh all-encompassing brand space'. Søren Askegaard (2006: 93) even argues that:

> in the face of growing competition in global markets and rising costs and clutter in mass-media advertising, leading to demands for efficiency, integrated communication and a search for alternative communication vehicles, the presence and importance of brands has arguably never been greater globally.

For practitioners working in the world of brands and branding, 'Brand is much more than a name or a logo. Brand is everything, and everything is brand' (Pallota 2011: 1) and 'there is no such thing as a world without brands' (Chevalier and Mazzolovo 2004: 3). Brand gurus, such as Wally Olins (2003: 7), see 'that what marketing, branding and all the rest of it are about is persuading, seducing and attempting to manipulate people into buying products and services. In companies that seduce, the brand is the focus of corporate life. Branding is everything.' For Naomi Klein (2000: 196) in her influential and popular political-economic critique, *No Logo*, brand consultancies have become the new 'brand factories, hammering out what is of true value: the idea, the lifestyle, the attitude. Brand builders are the new primary producers in our so-called knowledge economy.' In academic social-scientific accounts, brands are now seen to constitute 'a central feature of contemporary economic life' (Lury 2004: 27), branding is a 'core activity of capitalism' (Holt 2006a: 300), and their prevalence and importance in shaping the organization and dynamics of the economy in space and time signals 'a major change in the character of contemporary accumulation' (Hudson 2005: 68). Given such claims and views of the role and importance of brands and branding in economy, society, culture, polity and ecology, critical study of their geographies is overdue.

The missing geographies of brands and branding

Despite their dramatic rise, pervasiveness and importance, the ways in which the geographies of space and place are inescapably intertwined with brands and branding have been unevenly recognized and under-investigated. There are at least several reasons for this relative neglect. First, the field of brands and branding is longstanding but recently fashionable and increasingly crowded in the academic, practitioner and popular literatures. Despite differences in meaning and usage, a simple count of articles with 'brand' and/or 'branding' in their title published between 1969 and 2009 demonstrates the dramatic growth in academic research since the late 1990s (Figure 1.5). This research effort has proceeded across numerous disciplines including architecture (e.g. Klingman 2007), business studies (e.g. Buzzell *et al.* 1994), economics (e.g. Casson 1994), economic history (e.g. da Silva Lopes and Duguid 2010), geography (e.g. Pike 2011b), international relations (e.g. Anholt 2006), marketing (e.g. de Chernatony 2010; Holt 2006a), communication and media studies (e.g. Aronczyk and Powers 2010; Aronczyk 2013), planning (e.g. Ashworth and Voogd 1990), political science

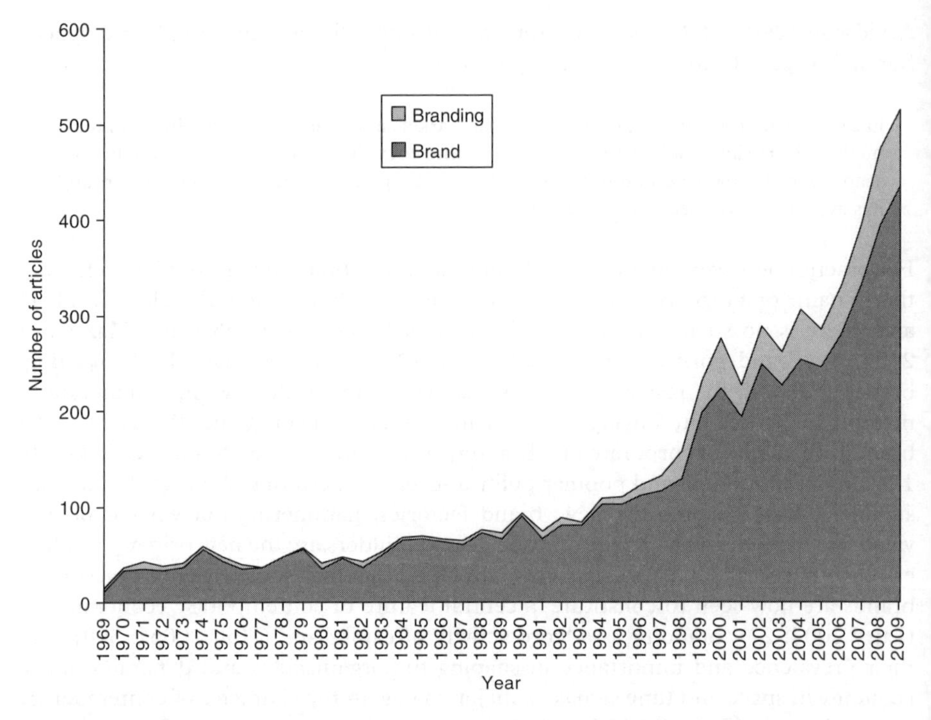

Figure 1.5 Number of articles with 'brand' and/or 'branding' in their title, 1969–2009. Source: Calculated from ISI Web of Knowledge data.

(e.g. van Ham 2008), tourism studies (e.g. Hankinson 2004), sociology (e.g. Arvidsson 2006; Lury 2004) and urban studies (e.g. Greenberg 2010; Hannigan 2004).

Echoing the importance of what Nigel Thrift (2005) termed the cultural circuit of capital and the soft infrastructure of knowledge creation and circulation, the work on brands and branding is undertaken too by a burgeoning industry generating a multitude of prescriptive guides and analytical frameworks from gurus and practitioners (e.g. Anholt 2006; Hart and Murphy 1998; Olins 2003) as well as academics who also provide services as consultants (e.g. de Chernatony 2010; Kapferer 2005). Texts range from influential and multi-edition analyses (e.g. David Aaker's (1996) *Building Strong Brands*) through current or former practitioner reflections (e.g. Saatchi and Saatchi's Kevin Roberts' (2005) *Lovemarks*) to more populist business advice accounts (e.g. Al and Laura Ries' (1998) *The 22 Immutable Laws of Branding*) and even self-help-style manuals (e.g. Tom Peters' (1999) *The Brand You 50*). Books about brands and branding have proven popular and regularly feature amongst the best-selling business books (Table 1.2).

International consultancy groups – such as Brand Finance, Futurebrand, Interbrand, Landor, Place Branding, Saffron and Wolff Olins – are also active knowledge producers. Their businesses focus upon developing and communicating proprietary branded services such as strategic advice and valuation methodologies, contact networks and

Table 1.2 Top five marketing books on branding, 2012

Title	*Author(s)*
1. *Emotional Branding: The New Paradigm for Connecting Brands to People*	Marc Gobé
2. *The 22 Immutable Laws of Branding*	Al and Laura Ries
3. *Unleashing the IdeaVirus*	Seth Godin
4. *Experiential Marketing: How to Get Customers to Sense, Feel, Think, Act, and Relate to Your Company and Brands*	Berndt H. Schmitt
5. *Building Strong Brands*	David A. Aaker

Source: Top 5 Marketing Books on Branding, http://marketing.about.com/od/brandstrategy/tp/top5branding.htm, accessed 14 November 2014.

commentary and analysis of the world of brands and branding (Aronczyk 2013; Moor 2008). As André Spicer (2010: 1736) points out:

> a remarkable amount of collective cognitive effort is committed to ruminating about brands. … There are a lot of people lurking in the lofts of our creative cities who devote their days to thinking about brands. The postmodern workforce is now glutted with brand workers who do everything from devising clever advertising campaigns to designing packaging or writing service scripts to be mouthed by bored teenagers working in a mall somewhere in nowheresville.

Specialized community building, networking events and media channels have blossomed too, including global conferences, web sites and community blogs organized by networks including *Brand Channel* and *Brand Republic*. Brands and branding are covered regularly in the wider business and financial press as well such as *Business Week*'s annual Top 100 Global Brands ranking produced jointly with Interbrand, *The Economist*'s (2009) edited collection on brands and branding and its periodic articles, surveys and futures pieces, and *The Financial Times* annual Global Brands Survey.

The world of brands and branding is, then, 'a young fledgling field … still in the making, on the move, influenced by agencies and consultancies as much as by scholarship and research. The boundary between truth, half-knowledge, common sense and sales talk is often hard to draw' (Kornberger 2010: 5). The increased production of knowledge about brands and branding can be divided between two broad, sometimes overlapping, camps. In one, exponents are focused on prescriptive work concerned with developing specific definitions, frameworks and methodologies for brands and branding, and advising commercial practitioners how to improve their effectiveness and impact. In the other, protagonists are engaged in more reflective and sometimes critical studies seeking to conceptualize, theorize and question the specific and wider purpose, value and effects of brands and branding. The diversity and variety in the approaches, purposes, sources and ways of thinking about brands and branding in these two broad camps have fostered only limited, partial and fragmented engagement with their geographies.

Second, the ways in which actors conceive of and use brands and branding have become more sophisticated in their interrelationships with goods and services commodities, complicating the task of interpreting their geographies. The traditional 'social engineering' paradigm in marketing from the 1950s has fragmented and been replaced by the growing sophistication and variety of branding strategies of brand owners and specialized consultants (Arvidsson 2006; Holt 2004). The initial 'product-plus-brand' approach has evolved into a wider and more holistic notion of 'brand-as-concept' (de Chernatony and McDonald 1998). In this perspective, actors frame brands as 'the tools used to detach "things" from the limited functionality of products and make them the engine of an endless desire for self-actualization and lifestyle' (Kornberger 2010: 9). Branding practices have been extended and deepened beyond specific products to encompass wider and interconnected ranges of individual goods and services brands. Actors have sought to appropriate value through the construction of meanings in brands. This technique is an attempt to forge longer lasting relationships to lifestyles and social identities that appeal to sophisticated, aesthetically aware and reflexive consumers, especially from affluent and elite social groups (Kornberger 2010; Urry 1995). This rise and intensification of branding during the 1990s heralded a closer interrelationship with brands because:

> Almost all accounts produced at this time saw brands as incorporating far more than simply a name, trademark and associated badge or logo, and assumed instead that brands should embody 'relationships', 'values' and 'feelings', to be expressed through an expanded range of 'executional elements' and 'visual indicators'. (Moor 2007: 6).

Growing saturation, competition and sophistication in especially advanced western consumer markets (Streeck 2012) coupled with the emergence of new forms of market research, consumer behaviour and media prompted the search for deeper and stronger brand attributes. This activity is focused upon constructing especially 'intangible ideals' (Holt 2006a: 299) that were not easily replicable or substitutable because 'differentiation in terms of function is less and less often able to sustain competitive advantage (because it can be imitated so quickly)' (Lury 2004: 28). The worth, visibility and burgeoning demand internationally for 'western' brands in emergent and faster growing economies and their nascent capitalist consumer societies further fuelled the logics of market segmentation, differentiated branding and the encouragement of brand literacy and loyalty (Ermann 2011). Brands and branding now pervade post-socialist transition economies such as Russia as well as emerging economies such as Brazil (Figure 1.6).

Last, brands and branding have extended their social and spatial reach beyond only goods and services commodities in the economy and culture more widely and deeply into society, polity and ecology. The world of brands and branding – to varying and uneven degrees – now encompasses architecture, art, associations, campaigns, charities, cities, clubs, communities, ethical and fair trade, events, exhibitions, festivals, internet domain names, knowledge, localities, nation states, organic food, people, places, political parties, prizes, regions, religions, social movements, spaces, sporting institutions, supranational entities, technologies and universities (Aronczyk and Powers 2010; Moor 2007; Pike 2011b; Van Ham 2001) (Table 1.3). Indeed, in the commercial sphere, it can

Figure 1.6 Brands and branding in Brasilia, Brazil, and Novosibirsk, Russia. Source: Author's images 2012 and 2013.

Table 1.3 The world of brands and branding

Sphere	*Examples*
Architecture	Norman Foster, Zaha Hadid
Art	Banksy, Rachel Whiteread
Associations	Girl Guides, Scouts
Campaigns	Occupy, UK Uncut
Charities	Oxfam, Red Cross
Cities	Be Berlin, **I am**sterdam
Clubs	FC Barcelona, Borussia Dortmund
Communities	Coin Street (London), The Eldonians (Liverpool)
Ethical trade	Traidcraft, The Fairtrade Foundation
Events	Olympics, Tour de France
Exhibitions	The Turner Prize, Future Generation Art Prize
Festivals	Glastonbury, Venice Film Festival
Internet domain names	Amazon.com, Patagonia.com
Knowledge	Clusters, Creative Class
Localities	The City of London, Wall Street
Nation states	Cool Britannia, Brand Singapore
Organic food	Organix, Whole Foods Market
People	Beyoncé, Brand Beckham
Political parties	New Labour, Lega Nord
Prizes	Nobel, Oscars
Regions	Catalonia, Third Italy
Religions	Catholicism, Scientology
Universities	Harvard University, Shanghai Jiao Tong University
Social movements	The WOMBLES, Indignados
Spaces	Motor Sport Valley, Silicon Valley
Sporting institutions	Boston Celtics, LA Lakers
Supranational entities	International Monetary Fund (IMF), Organisation for Economic Co-operation and Development (OECD)
Technologies	Microsoft Windows©, SAP©

at times appear 'as if there is hardly any market arena, not even a niche, that has been left uncolonized by branding processes' (Goldman and Papson 2006: 328). Brands and branding have spread and pervaded beyond the market as branded commercial interests and their logos touch and even saturate realms of economic, social, cultural, political and ecological life. Brands and branding generate powerful memes capable of transmitting ideas, practices and symbols globally. In one geographical imagination, the United States of America can be mapped by each state's most famous brands (Figure 1.7). Michel Chevalier and Gérald Mazzalovo (2004: 26) argue that:

> it shouldn't be surprising, in a society characterized by exponential growth of communication in all its forms and contents, that brands should be at the heart of contemporary

Figure 1.7 'The Corporate States of America'. Source: Steve Lovelace.

> life. They guide the purchases we make, influence our judgements about products and persons, and force us to position ourselves in relation to the values (or counter-values, or the absence of values) they communicate.

Digitization and media pluralization have deepened, widened and accelerated such processes. In the burgeoning software applications market, a popular app is called 'Logos Quiz' in which players are tested on how many brands they can recognize from incomplete brand logo images. 'Logos Quiz' ranked 67th on the top list of Apps and formerly occupied first position 3 months after its initial release.

In the polity, brands and their branded logos have become the foci for political dissent and resistance in relation to specific high profile brands and wider consumer society. Popular accounts agitating and articulating this agenda include Naomi Klein's (2000) *No Logo*, the 'culture jamming' of Kalle Lasn's *Adbusters* (2012) and its 'meme wars', and graffiti artist Banksy's 'brandalism' (Gough 2012). Even language and forms of communication in our private and public discourse that provide ordering devices to shape social thought and action in powerful ways have been infiltrated by brands and branding (see O'Neill 2011). Martin Kornberger (2010: 192) even asserts that 'lifestyle is our grammar, brands our alphabet'. Particular branded goods and services commodities have become adjectives (e.g. something is described using the food brand name 'Marmite' to mean people will either love it or hate it), nouns (e.g. the name given to something is synonymous with a brand name such as

'Rolls Royce' to convey something of the highest quality), verbs (e.g. an internet search becomes 'to Google') and metonyms (e.g. where a particular brand name becomes the descriptor and figure of speech for whole product categories such as Aspirin, Post-It and Tipp-ex). For brand owners, achieving such a place deep in the 'social lexicon' is something to which to aspire (Rigby 2012: 1).

The multi-faceted, pervasive and – it is argued here – *inescapably* spatial nature of brands and branding lie at the heart of the task of interpreting their geographies. Brands and branding intersect economic, social, cultural, ecological and political worlds. They are simultaneously: 'economic' as branded goods and services commodities in markets; 'social' as collectively produced, circulated and consumed objects; 'cultural' as entities providing meanings and identities; 'ecological' as material transformations, uses of nature and protective marks of assurance; and, 'political' as regulated intellectual properties, traded financial assets and contested symbols (Pike 2009a). Despite attracting growing attention across academic disciplines and propelling the burgeoning output in publications, the dramatic rise, pervasiveness and reach of brands and branding have made it difficult for research and analysis to keep pace.

Although a literature is emerging (Cook and Harrison 2003; Edensor and Kothari 2006; Jackson *et al.* 2007; Lewis *et al.* 2008; Pike 2009a, 2009b, 2011a, 2011b, 2013; Power and Hauge 2008; Power and Jansson 2011; Tokatli 2012a, 2013), the *geographies* of brands and branding have been relatively neglected. They are under-researched and lack conceptualization and theorization, analytical and methodological approaches, and a cumulative and substantive body of empirical research. Economic geography, for example, has 'consistently undervalued brands as an area of study' and 'many of our theories and accounts stop abruptly as the products leave the factory gates' (Power and Hauge 2008: 123, 139). Brands and branding geographies are overdue critical enquiry. This research need is recognized in marketing because, first, its focus has been shifting from 'Country of origin' to 'Country of origin of *brand*' (Phau and Prendergast 2000: 159, emphasis added). Second, the now well-established and 'ever growing use of origin identifiers by companies in marketing their products' (Papadopoulos 1993: 10) has been afforded more attention. Last, studies have sought to understand how and why consumers are becoming 'more interested in the origin of product ingredients and demand greater transparency over supply chain issues' (Beverland 2009: 189). Actors involved in brands and branding are deploying increasingly sophisticated strategies, frameworks, techniques and practices, drawing upon geographical associations to differentiate the meaning and value of goods and services commodities. Connections to spaces and places are being made to summon up distinctive and commercially beneficial attributes such as authenticity, quality, durability, style and cool in specific market settings.

The aims and organization of the book

Engaging the *inescapable* yet neglected geographies of brands and branding, this book aims to understand and explain what such geographical associations are, how and where they work, who creates and articulates them and what they mean for people and places. The central idea of origination is introduced and conceptualized. Origination means the attempts by actors – producers, circulators, consumers and regulators inter-related in spatial circuits – to construct geographical associations for goods and services commodities. Such geographical associations are used to connote, suggest and/or appeal to particular spatial references. They form part of the efforts of actors to create, cohere and stabilize meaning and value in specific brands and their branding in particular spatial and temporal market contexts. In the midst of more thoroughly branded contemporary capitalism, origination connects to what Karl Marx (1976: 165) interpreted as the attribution of mystical powers to the commodity form of material objects in the 'commodity fetish'. The argument addresses David Harvey's (1990: 422) call to 'get behind the veil, the fetishism of the market and the commodity, in order to tell the full story of social reproduction'. And it engages the 'de-fetishization' critique (e.g. Barnett *et al.* 2005). The conceptualization and theorization of origination offers a means of lifting what Miriam Greenberg (2008: 31) calls the 'mystical veils' woven around goods and services commodities by the increasingly advanced and sophisticated activities of actors for brands and branding. Their strategies, techniques and practices seek carefully to create, manage, rework and sometimes obscure the provenance of where goods are made and/or services are delivered from and – crucially – the economic, social, political, cultural and ecological conditions in which and where they are organized.

Origination makes its contributions in directly addressing the relative lack of research on the geographies of brands and branding. It provides the conceptualization and theorization capable of engaging the spatial dimensions of the brands and branding of goods and services commodities in more critical and geographically sensitive ways. The argument moves on from the narrow and simplistic interpretations of brands as carriers of homogeneity and uniformity in globalization and the limited geographies of the national frame of 'Country of Origin'. *Origination* seeks to advance geographical theory and stimulate work on the geographies of brands and branding in other disciplines in social science by forging and demonstrating the worth of fresh linkages at the intersection of political *and* cultural economy approaches to understanding and explaining spatial circuits of meaning and value. The distinctive contribution is the new theory of origination. It foregrounds and illuminates the roles of a broader range of actors including circulators and regulators rather than just the producers and consumers that constitute the focus of much contemporary work on brands and branding. Such actors are interrelated and animated by logics and rationales in spatial circuits. They try to work with geographical associations to create and fix meaning and value in and through branded goods and services commodities and their branding in particular market times and spaces.

Chapter 2 establishes the conceptual and theoretical foundations of the geographical reading of brands and branding. The argument builds upon the growing recognition but limited and underdeveloped grasp of the spatial aspects of brands and branding in especially marketing and sociology in social science (e.g. Arvidsson 2005; de Chernatony 2010; Holt 2006a). It builds upon and advances the emergent geographical accounts (e.g. Edensor and Kothari 2006; Lewis *et al.* 2008; Power and Hauge 2008). What is meant by brand is defined as an identifiable kind of good and/

or service commodity comprising differentiated characteristics. Branding is conceived as making meaning from the articulation and communication of the brand's attributes. The inescapable geographical connections and connotations of brands and branding are explained. Brands and branding are understood as geographically differentiated in their manifestation and circulation and interrelated with spatially uneven development. The concept of geographical association is introduced. It is defined as the characteristic elements – material, symbolic, discursive, visual – of the identifiable branded commodity and branding process that connect and/or connote particular 'geographical imaginaries' (Jackson 2002: 3). Differentiated meaning and value are created, circulated and valorized in spatial circuits through the agency of interrelated actors: producers, circulators, consumers and regulators. Each tries selectively to construct, cohere and stabilize geographical associations in commodity brands and their branding in commercially valuable and meaningful ways for particular spatial and temporal market settings.

Chapter 3 defines and explains what is meant by origination: how and why actors try to construct geographical associations in branded commodities and their branding in efforts to create, cohere and stabilize meaning and value for particular goods and services in the times and spaces of specific markets. Geographical origin(s) and provenance are discussed before questioning the limited geographies of 'Country of Origin' research and its recent transitions to incorporate 'Country of Origin of Brand' (Phau and Prendergast 2000) and a wider range of geographies. Rather than referring literally to any essential primary source or genuine form from which others are derived, origination is variegated. Actors attempt to originate branded commodities. They deploy strategies, frameworks, techniques and practices of branding to articulate and communicate meaningful and valuable geographical associations: producing, circulating, consuming and regulating 'geographical imaginaries' (Jackson 2002: 3) in diverse and varied ways. But such actors constitute and are compelled by rationales of accumulation, competition, differentiation and innovation. Such logics continually disrupt their spatial and temporal fixes of geographical associations, risking the collapse of the commercial appeal, coherence and competitiveness of their branded goods and services commodities in specific market times and spaces. Examples from clothing and tele-mediated services are used to explain how the agency of actors unfolds and territorial and relational originations are deployed. Except perhaps when wholly new market entrants, brands are not just empty receptacles nor is branding unscripted narrative through which actors can simply invent, write and insert originated content afresh. Brands and their branding are inescapably imbued by meanings and values in particular contexts. They are shaped by their specific histories and geographies that mark and colour their originations. Such legacies and traditions circumscribe the scope of actors' agency to varying degrees and in different ways. The methodology and analytical framework situates Douglas Holt's (2006b) historical-sociological branding genealogies in geographical context and builds upon Michael Watts' (2005: 534) commodity 'lives' approach. The socio-spatial biographies method enables analysis of origination across a range of interrelated and shifting territorial scales *and* relational networks in the geographies of the brands and branding. The extended case analyses that follow illuminate, ground and test the value of origination as a conceptual and theoretical framework in support of its wider explanatory claims. The discussion is about more than just documenting the particular and contingent detail of the geographies of specific branded commodities and their branding.

Chapter 4 examines the actors involved in constructing the 'local' geographical associations deployed in attempts to create and cohere meaning and value in the brand and branding of Newcastle Brown Ale (NBA). Demonstrating the value of socio-spatial biography method, the geo-historical origination of the brand lies in the construction of a particular 'local' drawn from the city of Newcastle upon Tyne and north east region of England. The brand producer, circulators, consumers and regulators created and appropriated meaning and value from its traditions and values in its urban and regional commercial heartland and subsequent national distribution in the UK. The regulatory device of the European Union Protected Geographical Indication was even used to fix its spatial attachment to the city of Newcastle upon Tyne. The distinctive 'local' origination proved only a temporary fix of meaning and value in the north east and UK. It was disrupted by shifts and segmentation in circulation and consumption in the alcoholic drinks market, shaped by consolidating brewing and outlet ownership, aggressive and targeted brand promotion, and innovation in new branded commodities. The brand management's search for new markets articulated with a growing market segment in America. NBA's management and importers changed its origination to the 'national' scale. It was reframed as 'Imported from England' to construct a premium meaning and value for its new college-educated, typically male and affluent younger consumers. Substantial sales growth in America raised questions about where NBA should be produced given its need for an 'imported' branding. Origination explains how what appear to be strong – and at times articulated by certain actors as intrinsic – geographical associations created in brands to particular places at specific scales can be temporary and lose their commercial meaning and value over time and space. Actors tried to reconstruct the origination of brands and their branding in new spatial and temporal market contexts through subtly different geographical associations, incorporating different scales and articulating them in new spatial circuits of meaning and value.

Chapter 5 examines efforts by actors involved in the brand and branding of Burberry to construct a national origination based upon the geographical imaginary of 'Britishness' to create and evoke meaning and value in the international fashion business. Burberry's specific socio-spatial history afforded pliable sources of discursive, material and symbolic assets valued by consumers and protected by regulators. These resources have enabled the brand owner, designers and managers to construct meaningful and valuable geographical associations based upon characteristic attributes of authenticity, quality and tradition. A particular version of 'Britishness' propelled Burberry's steady post-war growth but was undermined by its narrow product range, conservative consumers and weakly controlled circulation and regulation. Revitalizing the origination of the brand's 'Britishness' has been integral to Burberry's commercial modernization. The brand producers and circulators have successfully reworked its heritage assets and reconnected with emergent consumer tastes. This origination fix remains temporary in the context of internationalization, shifting markets and subcultural appropriation. Origination demonstrates that even as the brand's material geographical associations of production were being internationalized by its owners and managers *beyond* the specific national territory of Britain, the creative design, styling, detailing and advertising remain explicitly originated *in* and *with* the national frame of Britain and Britishness. As integral constituents of its meaning and value, Burberry's origination is produced, circulated, consumed and regulated in particular market times and spaces worldwide seemingly independent of the brand's material connections to particular production locations.

Chapter 6 examines the 'global' origination of Apple as an internationally pervasive, commercially successful and hugely influential brand. While some interpret Apple as a 'global brand' (Hollis 2010: 25), origination explains how the production, circulation, consumption and regulation of the 'global' in and through the brand and its branding are shaped by and constructed through meaningful and valuable geographical associations situated in the specific location of Silicon Valley – whether conceived of as a local, sub-regional or regional scale entity. Characteristic attributes and the reputation of the particular place of Cupertino, Santa Clara County, California in the heart of Silicon Valley provided discursive, material, symbolic and visual resources for actors' efforts to construct, cohere and stabilize Apple's meaning and value in specific market settings internationally. The brand's early growth was strongly geographically associated with the specific national territory of America. Disruption by competition, innovation and internationalization left Apple floundering commercially as its distinctive meaning and value were undermined. Renewal of the brand's characteristic and differentiated attributes – originated in the geographically associations of Silicon Valley – was undertaken following the return of its co-founder Steve Jobs in the mid-1990s. Commercial revitalization and global articulation of the brand was effected by the brand producer and circulators' new strategy, products and services, design, and value chain reorganization on a more internationalized basis. In the context of regulatory rules on country of origin and product labelling, the participant actors have had to experiment with origination more explicitly to reflect the spatial reorganization and internationalization of its functional operations. The '*Designed by Apple in California. Assembled in China*' origination enables Apple to sustain the meaning and value of its branded products and services, and reap the wide profit margins from its differentiated meaning and value, premium price *and* cost efficient production. Such origination remains a spatial and temporal fix, however, given ongoing competition, financialization, innovation and internationalization as well as regulatory rivalry over ownership and control of IPR, political and popular critique, and anti-brand and anti-Apple sentiment. Although an internationally prominent, influential and high profile brand across the world, the origination of Apple by the actors involved is not only and simply 'global'. Instead of being somehow 'placeless' and having no spatial connections in its apparent global ubiquity, origination of the brand's meaning and value in specific geographical associations in Silicon Valley have underpinned its most commercially successful periods of growth.

Chapter 7 examines origination and its relationships to the development of territories. Origination of brands and branding by the participant actors shapes geographical associations that pattern where specific economic activities take place. Origination influences the geographies of economy in the locations of investments, jobs, supply contracts for goods and services commodities, distribution networks, retail outlets, regulatory approvals and so forth across the economic landscape. Against a background of inter-territorial competition and efforts by actors to brand spaces and places, geographical associations have become a focus of attempts by actors involved in territorial development to capture the benefits of output, investment and jobs as well as capability and reputation in the wider global value chains. Origination can contribute positively to indigenous, endogenous and/or exogenous development strategies, demonstrated in the cases of Harris Tweed, Scotland, and the regional government in Castilla La Mancha, Spain. Limitations of especially strong originations in brands and branding are explained in the problems of overspecialization, dependency and lock-in, exemplified by Eastman Kodak, Rochester, New York State and – demonstrating the ambiguous nature of brands and branding in territorial development – Harris Tweed, Scotland.

Chapter 8 draws out the argument and contributions of *Origination* in providing a more sophisticated understanding and explanation of the geographies of commodity brands and branding. Origination illuminates where goods and services are from and why. It explains how actors in spatial circuits try to construct and articulate geographical associations in brands and branding to particular spaces and places. These activities are attempts to create, cohere, stabilize and appropriate meaning and value in spatial and temporal market contexts. The main contributions of *Origination* are, first, defining and conceptualizing geographical associations various kinds, extents and characters as integral to the geographies of brands and branding. Actors in spatial circuits utilize geographical associations in their efforts to create and fix meaning and value in branded goods and services commodities in the times and spaces of particular markets.

Second, geographical associations are inherently unstable and subject to disruption by the rationales of accumulation, competition, differentiation and innovation that relate and compel actors in spatial circuits of production, circulation, consumption and regulation. Brands and their branding can only ever provide temporary and conditional coherence, shape and form to the meaning and value of geographical associations. They are constantly buffeted and troubled in the changing times and spaces of spatial circuits and market contexts. Third, origination demonstrates how social and spatial inequalities are reproduced over time and space as actors seek to create, fix and manage geographical associations in brands and branding in geographically uneven ways. Brand and branding dynamics are underpinned by socio-spatial differentiation. Actors perpetuate and, in turn, are propelled by processes of accumulation, competition, differentiation and innovation to seek out, create, exploit and (re)produce economic and social disparities and inequalities over space and time. Countering the 'de-fetishization' critique, origination provides a means of lifting the 'mystical veils' (Greenberg 2008: 31) woven by brand and branding actors in spatial circuits.

Fourth, socio-spatial biography of branded goods and services commodities and their branding provides a methodological and analytical approach to the empirical research of origination. This framework identifies and engages the actors involved in spatial circuits and their attempts to deploy geographical associations to construct and cohere meaning and value in brands and branding in shifting spatial and temporal market settings. Fifth, origination furthers the intersection between political and cultural economy approaches within geography and broader social science. Origination enables closer integration of political economic interpretations of the dynamics and rationales of the creation, distribution and appropriation of economic value with accounts emphasizing the cultural construction of meaning and identity. Origination demonstrates how actors in spatial circuits work the meaning and value of geographical associations in goods and services commodities brands and their branding in territorial scales *and* relational networks. Last, origination encourages reflection upon the critical and normative issues in its politics. Origination focuses attention upon the limits of brands and branding and the question of 'what kind of brands and branding and for whom?' It addresses whether and how branded goods and services commodities could be geographically associated in more progressive and developmental ways for people and places by the actors involved. Origination seeks to promote dialogue and deliberation about brands and branding in territorial development in the context of material challenges to existing economic, social, cultural, political and ecological arrangements from climate change, financialization, resource shortages and social inequality.

Chapter Two
The Geographies of Brands and Branding

Introduction

This chapter establishes the geographical understanding of brands and branding at the heart of origination. It tackles the definitional and conceptual questions of what is meant by the terms brand and branding. The inescapably geographical connections and connotations of brands and branding are explained. The conceptualization of geographical associations is introduced and defined. The role of actors utilizing brands and branding in spatial circuits of meaning and value is set out.

Defining the brand and branding

While a pervasive entity and object of study, the precise meaning of brand is not entirely resolved. In literal definitions, to brand is to burn, label or mark as well as to place indelibly in the memory or stigmatize (Collins Concise Dictionary Plus 1989). Originating in pre-Roman livestock and pottery and mediaeval trades, brands were used to mark identifiable distinctions in property as proof of ownership, production and/or marks of infamy and notoriety (Room 1998) (Figure 2.1). Brands established differentiated and recognizable identities for goods and services commodities, especially skilled trades, in competition (Tregear 2003). Brands marked and articulated the reputation of craft labour 'whose prosperity depended on making a name for their goods as what we today would call a "brand label"' (Sennett 2006: 68). Brand names, signs and logos evolved to identify and articulate the character of goods and services

Origination: The Geographies of Brands and Branding, First Edition. Andy Pike.
© 2015 John Wiley & Sons, Ltd. Published 2015 by John Wiley & Sons, Ltd.

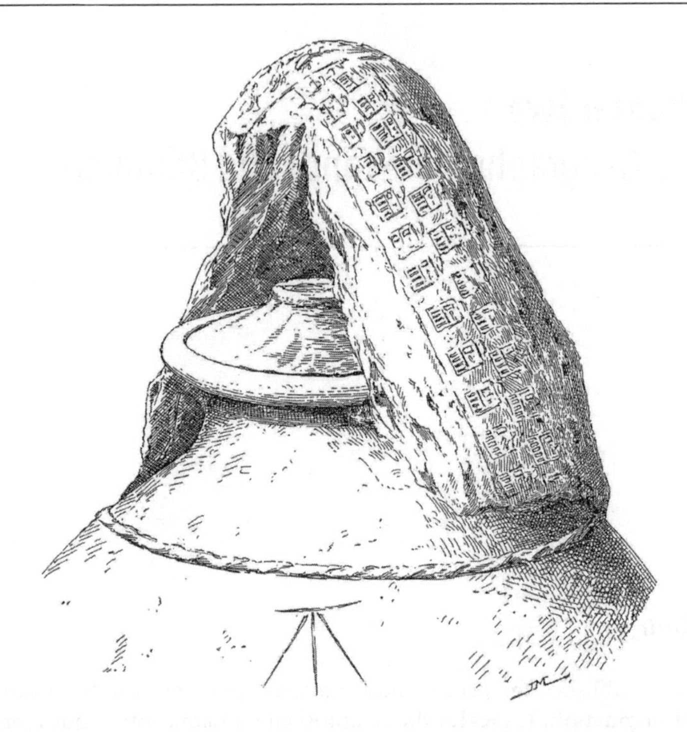

Figure 2.1 First Dynasty Egyptian wine jar, impressed with royal cylinder seal. Source: David Wengrow (2008: 10) from J. de Morgan (1897: 166) *Recherches sur les origines de l'Egypte*. Paris: Ernest Leroux (figure 527).

commodities (Riezebos 2003). They were used to reassure consumers of quality, provenance and geographical origin. The meaning *and* value of where goods and services commodities were from and who made and/or delivered them – what is understood and elaborated here as their origination – have been wrapped up in brands historically from their earliest emergence.

The rapid growth, evolving sophistication and widespread use of the term brand as a 'common currency' (Murphy 1998: 1) have been accompanied by an outpouring of definitions. Table 2.1 identifies one way of framing brands and branding in relation to the level of analysis and focus. The ways of conceiving of brands have proliferated. They include academic and prescriptive business, consulting and practitioner accounts competing to form opinion and provide – typically branded – proprietary advice and knowledge on a commercial basis (see, for example, Hart and Murphy 1998; Upshaw 1995). Multiple definitions and conceptual frameworks have emerged from the brand and branding 'industry':

> Which constantly churns out new terms around the topic with rather limited shelf lives, including 'brand equity', 'brand identity', 'brand strategy', 'brand image', 'brand reputation', 'brand promise', 'brand culture', 'brand experience', 'brand positioning', 'brand architecture' and 'brand awareness'. The word 'brand' seems to sell as soon as it is put in

Table 2.1 'The brand box'

	Focus on agency	*Focus on structure*
Organization and production as level of analysis	*Thesis*: Brands as management tool *Question*: How can we use the brand as a management tool next to other functions? *Key theorist*: David Aaker (e.g. 1996)	*Thesis*: Brands as corporate catalysts *Question*: How can we use the brand as a new paradigm to manage corporations? *Key theorists*: Mary Jo Hatch and Majken Schultz (e.g. 2008)
Society and consumption as level of anlysis	*Thesis*: Brands as sign *Question*: How do brands perform a signs, symbols and icons in society? *Key theorist*: Marcel Danesi (e.g. 2006)	*Thesis*: Brands as media *Question*: How do brands as a new interface restructure interaction between stakeholders? *Key theorist*: Celia Lury (e.g. 2004)

Source: Adapted from Kornberger (2010: 31).

> front of a more or less complex word. The resulting conceptual inflation does not help the clarity of the term. (Kornberger 2010: 15)

This activity has served to fragment rather than integrate understanding and explanation (Moor 2007).

An influential definition refers to the brand as the characteristic kind or variety of a particular good or service (de Chernatony 2010). No single or generally accepted 'one-size-fits-all' model of the tangible (e.g. design, function, quality) and, of growing significance, the intangible (e.g. feel, look, style) attributes of brands and their relative importance and relationships exists (de Chernatony and Dall'Olmo Riley 1998; Holt 2006a; Thakor and Kohli 1996). A refined and widely used conceptualization utilizes the concept of 'brand equity'. It is defined as the 'set of assets (and liabilities) linked to a brand's name and symbol that adds to (or subtracts from) the value provided by a product or service to a firm and/or to a firm's customers' (Aaker 1996: 7). In this framework, brand equity is a function of the closely connected assets of brand loyalty, awareness, perceived quality, associations and other proprietary resources that together generate meaning and value through their creation, articulation and enhancement by brand and branding actors (Table 2.2).

The brand equity framework provides some useful insights to address the tricky definitional questions. It emphasizes how the constituent elements of the brand's equity comprise a range of different assets both tangible and intangible, each of which are inescapably intertwined with geographical connections and connotations. The elements include associations (e.g. with particular people, periods and places), identities (e.g. image, look, style), origins (e.g. where it is designed, made, connected with or perceived to come from), qualities (e.g. feel, form, function) and values (e.g. efficiency, reliability, reputation). As Marcel Danesi (2006: 41) explains, as goods and services commodities become brands their 'use value is supplemented by a number of further associations'. Kevin Keller (2003) too interprets the meaning of

Table 2.2 Brand equity

Brand equity				
Brand loyalty	*Brand awareness*	*Perceived quality*	*Brand associations*	*Other proprietary brand assets*
Reduced marketing costs Trade leverage Attracting new customers Time to respond to competitive threats	Anchor to which other associations can be attached Familiarity-liking Signal of substance and commitment Brand to be considered	Reason-to-buy Differentiate and position Price Channel member interest Extensions	Help process and retrieve information Reason-to-buy Create positive attitude and feelings Extensions	Competitive advantage

Source: Adapted from Aaker (1996: 9).

the brand as fundamentally linked to its relationships with people, places, goods, services and other brands. The variety and pliability in the specific elements of brand equity underpin its appeal to actors seeking to construct and shape differentiated attributes and characteristics in commercially meaningful and valuable ways in particular spatial and temporal market contexts. The social and spatial histories of brands and branding can act as both meaningful and valuable assets *and* liabilities in specific market settings. The social *and* spatial approach to brand equity here addresses the critique of David Aaker's 'managerial' approach in its focus upon corporate brand ownership and control held by producers rather than consumers, its relative blindness to the social construction and consumption of brands, and the need to emphasize the relations between meaning and value (Kornberger 2010: 35).

Conceptualizing the brand as the object, branding can be conceived as the process of adding value to goods and services commodities by providing meaning (McCracken 1993). Branding ensures that goods and services are no longer just defined by their material basis and functionality but also by their 'symbolic powers and associations' (Kornberger 2010: 13). Branding emerged relatively recently and has become integrally related to brands (Arvidsson 2006). Branding too suffers from the same proliferation of competing new definitions and conceptualizations from commentators, practitioners and gurus that complicate the task of defining the brand (Moor 2007). Building upon David Aaker's (1996) 'brand equity' framework, branding can be seen as what actors do meaningfully to articulate, enhance and represent the facets and cues of the assets and liabilities in brands in ways that create value in certain market times and spaces (Moor 2007). Branding involves 'the non-material, creative side to production [that] relies heavily upon the input of signs and symbols to differentiate products and make them meaningful' (Allen 2002: 48). Martin Kornberger (2010: xii) interprets branding as the 'formula: brand = functionality + meaning. 3 M is innovation (not Scotch tape); Disney is entertainment (not just movies); Lexus is luxury (not just a means of transportation); Nike is performance (not just shoes).' This 'manufacture of meaning' (Jackson *et al.* 2011: 59) is longstanding as an advertising executive from Milwaukee put it in the 1960s 'The cosmetic manufacturers are not selling lanolin, they are selling hope. ...We no longer buy oranges, we buy vitality. We do not buy just an auto, we buy prestige' (quoted in Packard 1980: 35). Branding attempts to engender consumer trust and goodwill through constructing positive associations in the brand – such as authenticity, quality and style – that directly and positively influence purchasing decisions (de Chernatony 2010). Conceptualizing branding as a process of meaning-making encompasses the idea of '*re*-branding' for modifying or transforming the associations of brands. For example, efforts to reposition branded goods or services commodities in more appropriate and commercially promising market spaces and times (Dwyer and Jackson 2003).

Distinguishing the brand as the object and branding as the process is analytically helpful. But it is vital to recognize the integral nature of their relationship. The inter-dependency works through brand management. This practice seeks to align and coordinate the attributes, characteristics and values of brands through their broader circulation and promotion via various communication media through branding (Arvidsson 2006). In this way, brands are seen as performing the role of what Celia Lury (2004: 6) calls 'new media objects'. Branding is interpreted as a means of shaping, nurturing and valorizing the commitment and investments of consumers in brands in order to 'reproduce a distinctive brand image and strengthen brand equity' (Arvidsson 2005: 74).

Attempts by actors to construct the attributes and characteristics of the brand and its branding in meaningful and valuable ways is often highly selective. It is guided by particular interpretations and judgements about markets in specific spatial and temporal contexts, capturing and amplifying certain attributes and qualities of branded commodities while discarding and masking others. As practitioner Nigel Hollis (2010: 15) puts it: 'The last thing marketers can afford to do is to let brand associations pile up in an untidy heap'. In a sense, the brand is a composite of the facets and qualities of a good or service commodity deliberately chosen, integrated and communicated by actors as a coherent entity through branding. Except when wholly new market entrants, brands and their branding are not just empty receptacles into which owners can simply insert content afresh, however. Specific brands and their branding are inescapably imbued by meanings and values in particular settings, shaped by their histories and geographies.

The logic of differentiation is integral to the rapid rise and pervasiveness of brands and branding. This rationale is what Peter Jackson *et al.* (2011) describe as the construction of difference. Relative to commodified or more generic goods and services, the object of the brand and the process of branding are utilized by actors seeking product/image and price differentiation (Figure 2.2). In the late 1950s, the 'high apostle of image building', Pierre Marineau, explained to advertising executives that:

> Basically, what you are trying to do is create an illogical situation. You want the customer to fall in love with your product and have a profound brand loyalty when actually content may be very similar to hundreds of competing brands ... [the first task] ... is one of creating some differentiation in the mind – some individualization for the product which has a long list of competitors very close to it in content. (quoted in Packard 1980: 66)

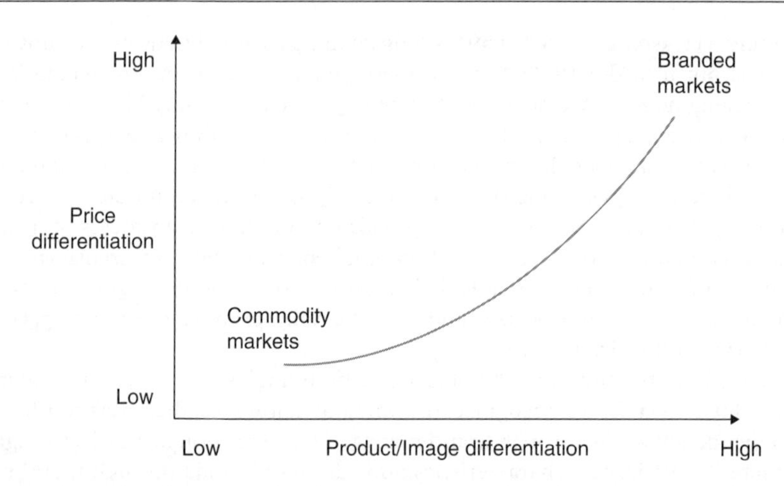

Figure 2.2 Price and product/image differentiation in commodity and branded markets. Source: Adapted from de Chernatony and McDonald (1998: 11).

This central logic has changed little since. Richard Sennett (2006: 143–4) describes this rationale as the search 'to make a basic product sold globally seem distinctive … to obscure homogeneity' so that 'The brand must seem to the consumer more than the thing itself'. Specific economic readings of brands and branding interpret this differentiation logic in different ways.

Under the assumptions of perfect information and competition, in neo-classical economics brands are considered unnecessary and wasteful: 'theoretically, in a perfect market, brands should not happen, period' (Kornberger 2010: xiii). If goods and services in markets are identical or substitutes then consumers are actually paying more for brand names because of the advertising and promotion costs incurred by producers and their premium pricing to increase profit margins. Given conditions of asymmetric and imperfect information, brands do perform an economic role. Brands provide information to consumers that they cannot easily verify themselves, lower search costs and provide guarantees of quality and reliability through confidence and trust-building (Casson 1994). They facilitate the free choices of individuals and exchange in markets by providing a signal to consumers through the price premium that brands provide high quality and reliability based on their past performance (Klein and Leffler 1981). Brands serve a specific purpose in relation to credence goods such as repair services, medical procedures, software programs and taxi journeys. These goods are marked by asymmetric information between buyers and sellers such that 'consumers can observe the utility they derive from the good *ex post*, they cannot judge whether the type or quality of the good they have received is the *ex ante* needed one' (Dulleck *et al.* 2010: 1). The brand reputation price premium is generated by advertising (Braithwaite 1928). It gets paid to reduce purchasing decision uncertainty by risk-averse consumers with imperfect information (Bauer 1960). Consumer awareness and recognition of brand names in markets incentivize firms to supply higher quality goods and services to enable higher prices to be charged and revenue streams to be maximized. Brands enable firms to increase their market share, create barriers to entry, enter new markets and reduce the price elasticity of goods and services relative to non-branded offers.

In his institutionalist analysis of the social context of consumption, Thorsten Veblen (1899) identified, first, status or Veblen goods for which demand increases rather than decreases with price because of the conferral of enhanced status upon the purchaser. Second, he explained positional goods that act as status symbols and signal their owner's relative position or standing within broader society. Both elements were integral to Veblen's (1899: iv, 111) theory of the 'conspicuous consumption of valuable goods'. Together, they were understood as 'a means of reputability to the gentleman of leisure', which are used to display income and status because 'As increased industrial efficiency makes it possible to procure the means of livelihood with less labor, the energies of the industrious members of the community are bent to the compassing of a higher result in conspicuous expenditure, rather than slackened to a more comfortable pace' (Veblen 1899: iv, 111). As powerful symbols of meaning and value, brands have become the shorthand for status and positional goods and services in contemporary society.

In a (neo-)Marxian approach, accumulation, competition and innovation dynamics propel differentiation by compelling actors to increase the surplus value yielded by the difference between perceived or exchange value and actual or use value (Hudson 2005). Brands and branding are part of capital's aim constantly to transcend wants and create new desires and needs that generate continuous demand to perpetuate the accumulation dynamic (Streeck 2012). Actors reinforce the growth in importance of the symbolic dimensions of use value to prevent commoditization through standardization, imitation and cost reduction driven by competition (Storper 1995). Contrary to Scott Lash and John Urry's (1994: 292) claims that they represent 'free floating signifiers' and 'sign values' emptied of their material content by the mobility and velocity of contemporary society (see also Du Gay and Pryke 2002), in (neo-)Marxian analysis brands are sign values capable of reaping symbolic rents by means of exercising reputational monopoly to appropriate (temporary) super-profits (Jessop 2008). This surplus is what David Harvey (2002: 94) terms 'monopoly rents'. Socially constructed images and identities in brands and branding underpin their differentiated meaning and value, appropriating geographical associations. This differentiation supports the price premium brands attract that 'represents what consumers are prepared to pay extra for the branded good in relation to other comparable goods. It represents the monetary value of the use-value of the brand' (Arvidsson 2005: 250).

As economic categories, brands have become functional and valorized. Corporate accounting practices and conventions incorporate brands as intangible assets on financial balance sheets. Brands influence corporate strategic planning and management as financial assets that materially affect their owners' share prices, access to capital and investor sentiment (Arvidsson 2005; Lindemann 2010). Brands have become financialized as securitized and tradeable assets. Brands produce cash flows repackaged into securities for sale to investors to raise capital (Pike and Pollard 2010; Willmott 2010). Brands have achieved the status of proven assets with predictable revenue streams. This has meant 'the securitization of intangible assets such as brands' has 'evolved into an established corporate financing tool' (Lindemann 2010: 87). Given their potential or actual value, brands have become central to corporate merger and acquisition activity. Indeed, the late 1980s and early 1990s takeovers of Rowntree by Nestlé for £2.8 billion (five times its book value) and Philip Morris by Kraft Foods for $12.9 billion (six times its book value) were the first deals in which the majority of the difference between market capitalization and purchase price was attributed to the value of the brands being acquired (Lindemann 2010; Willmott 2010). Brands accrue further rents from regulated intellectual property rights and trademarks through franchising, licensing and merchandising (Batchelor 1998). Royalties from licensing, for example, can be worth 15–20% of the sales price of the brand depending upon its industry. Critically, spatial and temporal context influences this valuation (Lindemann 2010).

Critical brand attributes and value relate to tangible characteristics such as function, physical appearance and quality. However, since many of their constituent elements are intangibles such as look or feel, the valuation of brands within existing accounting frameworks historically focused upon tangibles such as physical plant and stock has proved difficult, and competing methods have emerged (Lury and Moor 2010). A branch of international business consultancy has even developed around rival brand valuation methodologies (Willmott 2010). Interbrand's internationally influential proprietary method and widely publicized annual 'global' rankings, for example, illustrate the dominance of US-based brand owners such as Coca-Cola and Microsoft. In some cases, brand value can actually exceed the value of their owner's sales turnover and represent more than a third of their total market capitalization in the cases of Shell, Bank of America, Amazon, Coca-Cola and Google in 2005 (Table 2.3). Indeed, the financial value of the leading brands in some cases exceeds that of particular countries. Coca-Cola's $72 billion total brand value in 2011 places it at 64th position in the World Bank's 2011 national GDP rankings, between the Slovak Republic and Oman (Interbrand 2012; World Bank 2012). Commercial consultancy services such as Interbrand's – what Ngai-Ling Sum (2011: 165) calls 'knowledge brands' – are proprietary and branded ways of valuing the goodwill of specific brands. Significantly, such brands bear the strong imprints of geographical associations at the national level. These spatial connections are evident in the ways in which eligible brands are defined and how the spatially differentiated presence and power of brand owners in major markets is reflected and calculated.

Whether or not the premium price underpinned by differentiation matches consumer valuations in certain market times and spaces determines the commercial success or failure of specific brands. This is not just a rational or simple economic calculation. The economic value of any brand is inextricably intertwined with its cultural and social meaning. Branded goods and services commodities meet both functional *and* symbolic needs. To make them valuable in specific temporal and spatial market contexts, actors imbue brands with symbolic qualities and culturally endowed meanings through the meaning-making of branding. As Ray Hudson (2005: 69) puts it 'Purchasers thus pay for the brand name, the aesthetic meaning and cultural capital that this confers, rather than the use value of the commodity *per se*'.

Table 2.3 Interbrand/*Business Week* Ranking of 'Top Global Brands', 2005[a]

Rank	*Company*	*Brand value ($m)*	*Brand value as % of market capitalization*	*Brand value as % of total sales*	*Country of ownership*
1	Coca-Cola	67 525	64	290	United States
2	Microsoft	59 941	22	138	United States
3	IBM	53 376	44	54	United States
4	GE	46 996	12	28	United States
5	Intel	35 588	21	93	United States
6	Nokia	26 452	34	68	Finland
7	Disney	26 441	46	82	United States
8	McDonald's	26 014	71	128	United States
9	Toyota	24 837	19	14	Japan
10	Marlboro	21 189	15	22	United States
11	Mercedes	20 006	49	12	Germany
12	Citi	19 967	8	22	United States
13	Hewlett-Packard	18 866	29	22	United States
14	American Express	18 559	27	57	United States
15	Gillette	17 534	33	157	United States
16	BMW	17 126	61	31	Germany
17	Cisco	16 592	13	67	United States
18	Louis Vuitton	16 077	44	102	France
19	Honda	15 788	33	19	Japan
20	Samsung	14 956	19	26	South Korea

Note: [a]Nominal prices.
Source: Adapted from Interbrand (2005) Interbrand/*Business Week* Annual Ranking of the 100 'Top Global Brands.'

Brands and branding represent the valorization of the cultural forms and meanings of goods and services commodities (Scott 2000). Since the meaning of branded objects derives from their uses, forms and patterns of circulation (Appadurai 1986), Liz McFall (2002: 162) inteprets a contingent and – it is argued here geographically – associated notion of 'meaning … better understood as a contingent category constructed in instances of use and practice where the cultural and economic dimensions are not easily disentangled'.

Rather than just the latest set of fashionable, superficial and ultimately ephemeral marketing and advertising tools, brands and branding are integral to what Allen Scott (2007: 1466) terms the 'cognitive-cultural capitalism' of the contemporary era characterized by:

> marked intensification of competition (reinforced by globalization) in all spheres of the economy, though much of this competition occurs in modified Chamberlinian form because products with high quotients of cognitive-cultural content often possess quasi-monopoly features that make them imperfect substitutes for one another and hence susceptible to niche marketing strategies.

Having established the meaning of brands and branding, the accumulation, competition, differentiation and innovation rationales underpinning their emergence and connection of meaning *and* value, the next task in building the conceptual and theoretical foundations for origination is to specify exactly what makes brands and branding geographical.

The geographical in brands and branding

Inescapable geographical connections and connotations of branded goods and services commodities are longstanding and integral categories of brand and branding definition, value and meaning. But the ways in which such geographies of brands and branding are understood, conceptualized and theorized constitute a relatively neglected area. Harvey Molotch (2002: 665) claims this is because of the 'underappreciated ways that geographical space figures in making up goods'. And, it should be added, branded services commodities too. Brands and branding are geographical in at least three interrelated ways: inescapable geographical connections and connotations; geographically differentiated manifestation and circulation; and interrelationships with geographically uneven development (Pike 2009a, 2011a, 2011c, 2013). First, when conceptualized as an identifiable kind of good or service commodity, the brand is constituted of constituent characteristics imbued to varying degrees and in differing ways by spatial connections and connotations. As Dominic Power and Atle Hauge (2008: 138) argue, brands have an 'inherent spatiality'. Facets of brand equity – such as loyalty, awareness, perceived quality, attributes and associations – inextricably intertwine with spatially inflected considerations of who makes the good or delivers the service and from where as well as their identities, histories and socio-spatial connotations.

Actors involved in brands and branding construct spatial referents as part of efforts to create differentiated meaning and value. Noel Castree (2001: 1520–1) describes these spatial references as:

> The constructed imaginative geographies that are used to sell commodities via adverts, labels, trademarks, copyright or billboards ... [that] ... fill the vacuum of geographical ignorance with questionable, but commercially effective images of other places and cultures: think of Del Monte man, Uncle Ben's rice, or Jeep Cherokee.

Cultural economy approaches and recent studies of design, for example, demonstrate the strong connections between particular places and the differentiation of brands and branding. Certain territories become known as reputable sources of specific goods and services commodities (Scott 2000; Sunley *et al.* 2008). Places and brands intertwine to create valuable and meaningful associations, constructing and sharing cachet that seeps into space and/or place brands (Molotch 2005).

The emphasis upon spatial referents is not to suggest that the meaning and value in brands and branding is *only* constituted of geographically associated elements. Some characteristic attributes could have few or no spatial connections and connotations.

Or facets may have much weaker or perhaps meaningful but not valuable geographical resonance in specific spatial and temporal market settings. Examples include functional and/or intermediate goods and service commodities whose geographical associations are somehow hidden literally or discursively or that matter little to the meaning and value of the ultimate branded version of the good or service commodity and its branding. The brand and provenance of the structural steel in a building designed by a renowned architect such as Sir Norman Foster, the name and production location for a constituent component in a branded manufactured good or who processes your insurance claim and where may or may not be of consequence for the meaning and value of particular branded commodities in the spaces and times of certain markets, even though the rise of brands and branding in business-to-business markets demonstrates their growing reach (Arvidsson 2005).

Inescapable spatial ties and references unintended and/or undesirable can attach to brands in ways that make extrication difficult and cause problems of meaning and value for the actors involved. Rebellious consumer agency, for example, can contest and subvert the attempts of actors to control tightly the image, meaning and value of particular brands (Hebdige 1989). The Burberry brand's signature design was used by American rap music stars and lower class and poorer people or 'chavs' and down-market celebrities in the UK (Chapter 5). This appropriation generated concerns about this visibility and popularity corrupting the owner's version of the meaning and value of the brand, fuelling growth of counterfeit versions of Burberry branded merchandise in wider markets (Power and Hauge 2008; Moor 2007; Tokatli 2012a).

Branding is similarly entwined in geographical associations and contexts by actors in attempts to articulate the attributes and characteristics in brands in meaningful and valuable ways (Moor 2007). The meaning-making of branding is deployed by actors to identify, articulate and represent signs and symbols inescapably associated with the spatial context and connotations of brands. Branding practices including designs, logos and other symbolic tools are utilized to invoke and characterize often aspirational and spatially situated lifestyles: 'the branding dynamic uses place image to unite products and consumers who identify with a favored way of life and then sells them all elements of what it takes to live that imagined geographic life style' (Molotch 2002: 680). As Liz Moor (2007: 48) explains:

> Branding…is a kind of spatial extension and combination, in which previously discrete spaces of the brand – the advert, the point of purchase, the product in the home – are both multiplied, so that there are simply more 'brand spaces', and made to refer back and forth to one another so that they begin to connect up or overlap.

In this sense, branding can be interpreted as a spatially situated social practice (Thrift 1996), intimately connected with brands through brand management. Branding actors seek closely to control and shape the multiple and proliferating 'touch points' with the brand across time and space including mobile phones, in-store media, web sites, sponsored search links, social networks, word of mouth, viral videos, radio, TV and outdoor ads (Hollis 2010). Over time, branded goods and services commodities and their branding accumulate histories that are economic, social, cultural, ecological, political *and* spatial, and matter to their evolution. Such socio-spatial histories are integral to the conceptualization, theorization and analytical grasp of origination discussed in the next chapter.

Second, brands and branding are geographical because of the ways in which they (re)produce geographical differentiation across space and time. Modifying Michael Watts' (2005: 527) argument for the more thoroughly branded world of contemporary cognitive-cultural capitalism: 'The life of the [*branded*] commodity typically involves movement through space and time, during which it adds values and meanings of various forms. [*Branded*] Commodities are therefore pre-eminently geographical objects' (see also Smith and Bridge 2003; Smith *et al.* 2002). This role of brands and branding in geographical differentiation has fostered different explanatory perspectives. Some see ostensibly 'global' brands and branding as somehow placeless vehicles of globalization that cross borders as a 'global fluid' and are 'super-territorial and super-organic, floating free' (Urry 2003: 60, 68). Martin Kornberger (2010: 231) even argues that 'we experience globalization through brands: without brands globalization would be meaningless'. Such interpretations echo marketing visionary Theodore Levitt's (1983: 100, 92–3) influential 'earth is flat' argument caused by the globalization of markets in which 'Global competition spells the end of domestic territoriality'. Seeing the reach of global corporations through their brands, Theodore Levitt (1983: 93) claimed that:

> The global corporation operates with resolute constancy – at low relative cost – as if the entire world (or major regions of it) were a single entity; it sells the same things in the same way everywhere. ... The world's needs and desires have been irrevocably homogenized. ... Commercially, nothing confirms this as much as the success of McDonald's from the Champs Elysées to the Ginza, of Coca-Cola in Bahrain and Pepsi-Cola in Moscow, and of rock music, Greek salad, Hollywood movies, Revlon cosmetics, Sony televisions, and Levi jeans everywhere. 'High touch' products are as ubiquitous as high-tech.

The geographies of brands and branding in some of these accounts are marked by homogenization and uniformity (Ohmae 1992) and even 'sameness' (Wortzel 1987). A brand like Visa appears 'to have no provenance' and 'like many brands ... has global reach. National boundaries mean nothing to such brands. ... Brands like Mercedes can sweep across the world. Their physical and emotional presence is ubiquitous, and they seem omnipresent, almost omnipotent' (Olins 2003: 17). In this kind of interpretation, 'Country of Origin' has been rendered an 'irrelevant construct' (Phau and Prendergast 1999: 71). Focused upon brand not place and facilitated by information and communication technologies, 'brand communities' are emerging around brands as a 'specialized, non-geographically-bound community, based on a structured set of social relationships among admirers of the brand' (Muñiz and O'Guinn 2001: 412). Ubiquity and mobility of branded goods and services commodities are said to be turning spaces and places into what Anna Klingman (2007: 3) calls uniform commercial 'brandscapes' dominated by the same global brands and their branding of images, logos and signs. As Jonathan Schroeder and Miriam Salzer-Mörling (2006: 10) put it: 'the cultural landscape has been profoundly transformed into a commercial brandscape in which the production and consumption of signs rivals the production and consumption of physical products'. Echoing Thomas Friedman's (2006: 206) version of the 'flat world', such changes are seen in some accounts as lessening spatial differences in a flat and slippery world.

Qualifying and challenging such views is a more spatially sensitive interpretation of heterogeneity, diversity and variety in how brands and branding (re)produce geographical differentiation in a spiky and sticky world (See, for example, Hollis 2010; Jackson *et al.* 2007; Lewis *et al.* 2008; Pike 2009a, 2011a; Power and Hauge 2008; Quelch and Jocz 2012; see also Christopherson *et al.* 2008). In these conceptions, branded commodities travel and communicate differing meanings and values across space and time. Particular brands and their branding find geographically differentiated kinds and degrees of commercial, social, cultural, ecological and political meaning and value. In seeking to shape and respond to the particularities of different geographical and temporal market contexts, branding practices may similarly be spatially attenuated and heterogeneous. Economic anthropologists have shown how even ostensibly 'global' brands are appropriated and modified in particular geographical settings. A key example is Daniel Miller's (1998) study of the 'hybridization' of Coca-Cola in Trinidad. More geographically sensitive research in marketing has demonstrated the importance of managing the global attributes of brands in their adaptation and fine-tuning to exploit particular market places (Holt *et al.* 2004). Practitioners too acknowledge that 'few brands succeed in creating a strong connection with consumers across multiple countries. The formula that makes a brand strong in one country may not travel well. Consumer needs and value still differ dramatically from place to place' (Hollis 2010: 1). Martin Kornberger (2010: 26) cites the example of Deloitte business services in which 'the brand becomes a kit of parts that people can use and appropriate locally like a resource-box. The prefabricated tools ensure some homogeneity while the local *bricolages* allow for alignment with local context' (original author's emphasis). Critically, in this view:

> Brand identity is not based on sameness but on difference; brands are engines for exploiting differences; they absorb differences – and the local is an inexhaustible source of idiosyncracies. ... Brands do not dominate the world by unifying it: they rule through diversity and difference. To homogenize would represent a system failure ... different people interpret brands differently ... brands become different things for different people. (Kornberger 2010: 233)

Even when internationalized, for brands 'the territory from which they emanate ... as well as the territories to which they are exported remains of central importance' (Ritzer 1998: 12). Attributes attached to brands such as 'westernness', 'Americanness' or 'exoticness' are important in understanding their worldwide proliferation (Askegaard 2006). Geographical differentiation, then, is integral to the different ways in which different people in different places see, interpret and act in relation to the meaning and value of branded goods and services commodities and branding processes.

Third, brands and branding are geographical because of the close interrelationships between their inescapable and spatially differentiated geographies and 'uneven geographical development' (Harvey 1990: 432). Brands and branding depend upon and are constitutive of the spatial differentiation of economy and society. This reliance is because the actors involved are compelled by the rationales of accumulation, competition, differentiation and innovation to search for, create, exploit and (re)produce economic and social disparities and inequalities over space and time. Actors seek explicitly to construct and define markets for brands in time and space through their brand building and branding practices. Luxury brands, especially, focus on segmentation of markets where 'The brand aims at an urban public who have a high income level and are fond of traveling. It relies on these strong determinisms to eclipse geographical specificities' (Chevalier and Mazzalovo 2004: 127). Actors compete over the structure of markets. They look to innovate in a disruptive, Schumpeterian sense in pursuit of monopoly rents rather than just accepting the constraints of existing patterns of market categories (Harvey 2006; Hudson 2005; Slater 2002). In common with brands and branding, markets and their constituent structures are inescapably social *and* spatial; articulating, reflecting and penetrating socio-spatial relations between people and places (Peck 2012).

Brand owners are compelled by accumulation, competition, differentiation and innovation rationales to segment, defend and exploit profitable parts of goods and services commodity markets in particular spatial and temporal contexts. Accumulation informs directly the role of brands within marketing strategies. This logic was encapsulated with some clarity by Tony Palmer, Chief Marketing Officer, Kimberly-Clark: 'we have a very simple perspective on the role of marketing. It is to sell more stuff to more people, for more money, more often' (quoted in Hollis 2010: 205). Branding guru Wally Olins (2003: 10) too sees that 'Commercial brands exist because they are a powerful tool to help companies make money'. Brand actors invest much time, effort and resources grappling with social and geographical differences and specifically how they can be used and perpetuated to create and realize meaning and value in spatial circuits. Naomi Klein (2000: 117, 130) argued that brand producers desire 'a One World placelessness, a global mall in which corporations are able to sell a single product in numerous countries without triggering the old cries of "Coca-Colonization"' and 'market-driven globalization doesn't want diversity; quite the opposite. Its enemies are national habits, local brands and distinctive regional tastes.' But this scale imperative is only part of the story and is balanced and tempered by the rationales of differentiation and segmentation. Brand and branding actors generate and profit from unequal spatial differentiation. They use premium pricing to differentiate branded commodities relative to more commodified goods and services providing the same (or similar) use values and fulfilling the same needs. They construct aspirational spaces as profitable premium niches to tempt consumers to trade-up to perhaps less affordable offerings with higher profit margins (Frank 2000). They introduce cycles of fashion and season deliberately to quicken capital circulation (Harvey 1989). As Ray Hudson (2005: 69) explains, branding seeks to destabilize existing markets and re-institutionalize them around new, strategically calculated good and service brand definitions such that the 'aesthetic and cultural meanings of brands and sub-brands then become ways of segmenting markets by ability to make the premium payments required to possess the desired brand'. Such consumption capacity is differentiated socially and spatially.

In turn, spatial manifestations of economic and social differences and inequalities fuel market construction and segmentation because 'Wide disparities between rich and poor ... bring into being more luxurious types of goods than would otherwise exist' (Molotch 2002: 682). Social stratification and hierarchy underpin the relentless search by elite groups for distinction through consumption, propelling the continuous refinement and construction of new desires and wants beyond needs (Frank 2000; Streeck 2012). Brand and branding's differentiation imperative compels actors actively to seek out, perpetuate and (re)produce such inequalities, fostering social polarization since 'The new poor, without the right labels and brands, are not just excluded but invisible' (Lawson 2006: 31). In consumer-oriented society,

> Esteem and its opposite – stigma – are earned through consumption practices. Social outcasts are not the unemployed – those who are not productive – but those who do not or cannot consume. Poor is someone who cannot afford brand and has to settle for no-name home brands. (Kornberger 2010: 211)

The identification, reflection and orchestration of socio-spatial disparities – finding and tapping into the spatially uneven geographies of economy and society – are central to brand owner strategy. The Global Brand Director for Mars confectionary, for example, claimed 'the age of the average is dead' and sought sub-national 'pockets of affluence' as a branding priority (Murray 1998: 140).

Brands and branding perpetuate uneven geographical development, then, through what Celia Lury (2004: 37) explains as a:

> Hierarchical division of labour ... with design-intensive producers located at the top ... and many of those actually involved in manufacturing the products or delivering the service at the bottom ...only a few pennies of the price of a Starbucks cappuccino goes to pay for the labour of those who harvest and roast coffee beans, and not many more are paid to those who serve the drinks. The remainder accrues to those able to assert the value of their contribution to the brand in terms of creativity, product innovation or design activity.

In Allen Scott's (2007: 1468) contemporary 'cognitive-cultural economy', this hierarchical division of labour is social *and* spatial. It is (re)produced by the actors animating brands and branding in spatial circuits of meaning and value (Hudson 2008; Smith *et al.* 2002). Geographical unevenness appears inscribed in contemporary patterns of socio-spatial development. It encompasses the upper tier of elite occupations in metropolitan centres deploying 'heavy doses of the human touch ... for the purposes of management, research, information gathering and synthesis, communication, inter-personal exchange, design, the infusion of sentiment, feeling and symbolic content into final products' and the lower tiers 'employed in a thick stratum of manual production activities that are not as well paid and much less gratifying in their psychic rewards' (Scott 2007: 1468).

Inescapably spatial brands and branding perpetuate and reinforce geographically uneven development by forging and even amplifying unequal socio-spatial divisions of labour and competitive socio-spatial relations between spaces and places (Pike 2009a). Brand owners engage in the 'race to the bottom' of regulatory arbitrage in the outsourcing and exploitation of marginal labour pools internationally. Rival producers and circulators of competing brands from different places exacerbate inter-territorial competition (Pike 2011c; Ross 2004). Sometimes intersecting through branded goods and services markets, development institutions are increasingly involved in attempting to render their place distinctive relative to other places in the territorial competition for businesses, investment, jobs, residents, students, visitors and spectacle events (Hannigan 2004; Pike 2011c; Richardson 2012; Turok 2009) (Chapter 7).

The relationship between brands and branding and uneven development is central to the politics of commodity geographies. It connects to David Harvey's (1990: 422, 432) call to 'get behind the veil, the fetishism of the market and the commodity, in order to tell the full story of social reproduction' by 'tracing back' the relationships between commodities and 'uneven geographical development'. Uncovering the 'hidden life to commodities' can 'reveal profound insights into the entire edifice – the society, the culture, the political economy – of commodity producing systems' (Watts 2005: 533). Such 'de-fetishization' arguments have been subject to critique, however. Issues concern: the 'double fetish' of the thing-like quality of social relations embodied in commodities and the geographical imaginaries constructed by commercial interests (Cook and Crang 1996); the complexity of imaginative geographies in the social lives of commodities (Castree 2001); the privileging of academic over popular, increasingly sophisticated consumer knowledges (Jackson 1999); the imagining of alternative geographies by 'working with the fetish' (Smith and Bridge 2003: 262); and, the questioning of commodity fetishism's relevance and whether 'reconnecting' will 'restore to view a previously hidden chain of commitments and responsibilities' (Barnett *et al.* 2005: 24). The contribution of origination to such concerns is returned to in Chapter 8.

Geographical associations in brands and branding

Having established the geographical in brands and branding, the next task is to conceptualize and theorize how, why, in what ways and by whom geographies are inescapably entwined in branded goods and services commodities and their branding. A helpful reading draws from the debate about entanglement in economic anthropology and economic sociology. This dialogue focuses upon the ways commercial imperatives in markets are driving towards the ever more inclusive entanglement of the 'transactable object' of goods and services in the life of consumers (Barry and Slater 2002: 183). Actors are interpreted as seeking to make commodities more meaningful and resonant across a 'diversity of values and value systems' and a range of registers including the rational, aesthetic, cultural and moral (Barry and Slater 2002: 183). Competition is seen as driving innovation and differentiation through the 'qualification' or attribution of qualities and 'singularization' or emphasis upon difference and uniqueness of goods and services commodities and their closer and more enduring attachment to consumers (Callon 2005: 6). The debate revolves around how this process of entanglement unfolds.

Michel Callon (*et al.* 2002, 2005) interprets a necessary moment of framing and *disentanglement* through market transactions to enable the participants in exchange to free themselves from further and ongoing ties that might prevent the transfer of ownership and property rights over goods and services commodities. In contrast, Daniel Miller (2002: 227) sees an ongoing process of increasing *entanglement* because 'most industries have to engage in highly qualitative and *entangled* judgements about looks and style and image and "feel" out of which they may, if they have the right sense of the "street", make a profit. *The way to profitability is not through disentanglement, but through further entanglement*' (emphasis added).

In a geographical reading of this debate, Roger Lee (2006: 422) supports Daniel Miller's view of entanglement in the inseparability of economy/society because the:

> Entangled economic geographies ... remain unframed – or rather multiply framed – in the senses both that the agents, objects, goods and merchandise involved in them remain more, or less, imperfectly distinguished and associated with one another and that multiple social relations are at play between them.

Roger Lee (2006: 414) interprets an economism in Michel Callon's view and a desire for a 'purification of economic relations' that risks missing 'the inherent complexity of ordinary economies' and effectively 'places limits on the economic geographical imagination'. This debate is informed by a post-structuralist ontology that tends to give insufficient recognition to the geographical political economies of the rationales of accumulation, competition, differentiation and innovation, spatial circuits of capital and the regulation of the state and other institutions. However, the dialogue about entanglement echoes important elements of the definition, conception and explanation of the geographical in brands and branding discussed above. The protagonists acknowledge the rise of brands and branding as devices for differentiation, understood as 'qualification' and 'singularization'. They recognize too, especially Daniel Miller, the emergence of 'brand-as-concept' reaching beyond specific commodities through the deployment of the meaning-making of branding by actors trying to bring goods and services into deeper, more resonant, longer lasting and more profitable relationships with consumers. This phenomenon is what Kevin Roberts (2005: 1) terms 'loyalty beyond reason'. Conceptualizing always and ongoing *geographical* entanglement underpins the understanding articulated here of the *inescapable geographical associations* of branded commodities as 'transactable objects' and the meaning-making of branding.

The geographical in brands and branding can be conceptualized as geographical associations inextricably entwined in the value and meaning of goods and services commodities. Ian Cook and Philip Crang (1996: 132) have demonstrated how 'geographical knowledges – based in the cultural meanings of places and spaces ... [are] deployed in order to 're-enchant' ... commodities and to differentiate them from the devalued functionality and homogeneity of standardized products and places'. As a concept with layers of meaning (Harvey 1996) and 'cultural resonances' (Scott 2010: 122), place is particularly amenable to forms of social construction. Through this process Peter Jackson's (2002: 3) 'geographical imaginaries' are appropriated and

Table 2.4 Geographical associations in brands and branding

Kind	*Extent*	*Nature*
Material	Strong	Inherent
Discursive	Weak	Authentic
Symbolic		Fictitious
Visual		Hybrid
	Examples	
'Made in...'	Name	Geographical indications
Language	Design	Labelling
Logo	'Placeless' ubiquity	Styling
Place images		

connoted by actors *in* and *through* geographical associations in brands and branding processes. Wider social science accounts recognize and support the idea of geographical association in goods and services commodities in 'spatial identifications' (Miller 1998: 185) in economic anthropology, 'country and cultural signifiers' (Phau and Prendergast 2000: 164) in marketing and in how 'place gets into goods by the way its elements manage to combine' (Molotch 2002: 686) in sociology. Indeed, sociologist Adam Arvidsson (2005: 239) argues that 'Building brand equity is about fostering a number of possible attachments around the brand ... experiences, emotions, attitudes, lifestyles or, most importantly perhaps, loyalty'. Each of these attachments intersect and articulate geographical connections and connotations of association.

Geographical associations provide the conceptualization through which to understand and explain the geographical in brands and branding: the inescapable spatial connections and connotations; the geographically differentiated expression; and, the interrelation with socio-spatial inequalities. Geographical associations are the characteristic elements of the identifiable branded commodity and branding process that connect and/or connote particular 'geographical imaginaries' (Jackson 2002: 3). They are more than just the material ties of brands and branding in fixed relationships to certain spaces and places. Geographical associations are of different kinds, varying in their extents and natures over space and time (Pike 2009a) (Table 2.4).

The different kinds of geographical associations are multiple and overlapping including material, symbolic, discursive, visual and aural. Material geographical associations might include specific spatial connections to authentic and traditional methods and particular places of the brand's production (Dwyer and Jackson 2003). Direct geographical references are made in a brand's name, for example, such as Aberdeen Angus Beef and Welsh Lamb (Morgan *et al.* 2006). Material geographical associations are manifest in tangible attributes of brands such as taste, fabric or user-friendliness. Symbolic geographical associations imbue and insinuate spatial referents in brand logos as proprietary markers circulated to draw attention and custom from potential consumers. Aspects of place geographically associated in brands and their branding are part of internationally legible, meaningful and valuable visual languages, exaggerating place-themes by 'using materials and designs that connote ... favored geographic spots' (Molotch 2002: 678). Discursive geographical associations attempt to align brands with aspirational and desirable spaces and places through stories and narratives in print and on-line advertising as well as fiction and non-fiction in literature. Guinness, for example, plays upon its particular Irish history and literary tradition (Griffiths 2004). Visual geographical associations utilize what Mrugank Thakor and Chiranjeev Kohli (1996: 33) call 'origin images ... recalling to consumers a rich set of associations' to surround and imbue brands and infuse branding concepts and messages. New York City is appropriated by numerous brands, such 'NYC' cosmetics, as a shorthand to convey chic, modernity and urbanity. Aural associations signify geographical connotations through meaningful music, songs, poetry, language, slang, accents and dialects. Sportswear brand Nike's use of distinctively Brazilian Samba music and players from the Brazilian national team in adverts, for example, explicitly connects its branded sportswear with Brazil and its illustrious sporting history (Haig 2004a).

Different kinds of geographical associations can connect and overlap in the brand and branding practices of the actors involved. Spatial alignment situates the geographical associations of a brand in a specific brand name and particular territorial locality such as 'Tom's of Maine' (Molotch 2002). Spatial discontinuity is evident where geographical associations are selectively obscured. Fashion clothing and footwear brands trading and branded as 'Made in Italy', for example, are increasingly using but obscuring their use of international production outsourcing to lower cost locations (Hadjimichalis 2006; see also Thomas 2007, 2008; and Ross 2004).

Geographical associations in branded goods and services commodities encompass differing degrees and varied natures of attachment, connection and connotation. The quantitative and qualitative extent and nature of geographical associations illuminate understanding of how they constitute and relate to branded objects and branding processes. Geographical associations of different degrees and characters shape the agency of producers, circulators, consumers and regulators in their attempts to construct and stabilize spatial connections and connotations in branded goods and services. An array of distinctions can be postulated not as simple binary opposites or static descriptors but as starting points in framing conceptualization. Table 2.5 offers a preliminary but not exhaustive vocabulary to capture both territorial scalar and relational network dimensions of geographical associations. For example, geographical associations in a brand and its branding can range between situations where the meaning and value are agreed or contested amongst actors in its spatial circuit. Geographical associations can vary from being spatially bounded and limited to a specific territory to being spatially unbounded and circulating in relational networks. Table 2.6 develops the conceptualization further to identify particular themes, characteristics, practices and elements through which geographical associations are manifest in brands and branding. Taking the specific theme of economy, for example, a number of characteristics can be invoked such as quality, tradition and reputation. Actors work these and other attributes into their brand and branding practices and elements through design, name and labelling. Scottish Widows, for instance, articulates the frugality, integrity, prudence and trustworthiness geographically associated with the national territory of Scotland in its branded services. These values are symbolized in the brand name and

Table 2.5 Distinctions in geographical associations

Distinctions
Agreed – Contested
Deep – Shallow
Extensive – Limited
Fast – Slow
Fixed – Fluid
Hard – Soft
Homogenous – Heterogenous
Intrinsic – Instrumental
Material – Symbolic
Permanent – Temporary
Spatially bounded – Spatially unbounded
Strong – Weak
Thick – Thin
Tight – Loose
Transparent – Opaque

advertising and are material in the functional financial products they sell to consumers such as bank accounts and pensions.

Actors articulating branded goods and services commodities can try to articulate strong and authentic material associations to specific geographical origins. In agro-food brands especially, intrinsic ties are evident in the biophysical connections between branded commodities and particular places such as Champagne wines and Parma ham (Morgan *et al.* 2006). Strong, deep and resonant geographical associations are then used to script brand content and branding activities around particular mores and tropes. Brands inherently connected and synonymous with certain places enable actors to draw upon their particularity. This specificity is sometimes encapsulated in their brand name such as Newcastle Brown Ale (see Chapter 4), *Parmigiano Reggiano* cheese and Pierre Cardin Paris clothing and accessories. Such brands are in a sense 'region-bound' by their close, tight and clearly identified geographical associations and 'take on their point of origin almost as a defining attribute' (Molotch 2002: 672, 677). While lacking inherent geographical associations that mean such goods and services commodities could not technically be produced or delivered from elsewhere, actors seek commercial advantage through constructing explicit spatial connections in brands. Nordic brands of extreme sports equipment such as clothing, skis and snowboards from Helly Hansen and Klättermusen, for example, articulate meaning and value by demonstrating their elite specification and high quality. The brand actors emphasize their design, testing and development in the harshest environments in northern Europe (Hauge 2011). Other actors might embody weaker and constructed geographical associations to entirely fictitious places branded to appeal to particular social and spatial market segments at certain times. Peter Jackson *et al.* (2011: 59) demonstrate this process of 'manufacturing meaning' in a UK food

Table 2.6 Themes, characteristics, practices, elements and brand and branding examples of the geographical associations of brands and branding

Themes	*Characteristics*	*Practices and elements*	*Examples of brand and branding and geographical associations (brand owner, HQ location)*
Economy	Efficiency Quality Reputation Tradition Value Workforce/skills	Design Name Labelling Packaging	*Sony* (electronics) – Japanese ingenuity, high-technology and innovation (Sony, Tokyo, Japan) *Scottish Widows* (financial services) – Scottish frugality, integrity and trustworthiness (Lloyds TSB, London, UK)
Society	Architecture Ethnicity History Language Reliability Style Trust Values	Colour Design Image Logo Name Presentation	*BMW* (automobiles) – German rationality, technical sophistication and reliable engineering (BMW Group, Munich, Germany) *IKEA* (furniture and fittings) – Scandanavian design, style and minimalism (Inter IKEA Systems, Delft, Netherlands)
Polity	Administrations Charisma Competence Institutions Political leaders/parties Traditions Vision	Crests Emblems Flags Images Names Packaging Symbols	*Swissair* (airline) – Swiss efficiency, neutrality, quality and reliability (Swissair, Zürich, Switzerland) *Coutts* (private banking) – Discretion, status, British integrity and tradition (Royal Bank of Scotland, Edinburgh, UK)
Culture	Artefacts Folklore Icons Identity Myths Texts Traditions	Design Images Logos Packaging Styling Variety	*Prada* (fashion) – 'Made in Italy' design, style and quality (Milan, Italy) *Quicksilver* (Surfing clothing and equipment) – Hip and laid-back beach style (Torquay, Australia)
Ecology	Authenticity Intrinsic attributes (e.g. smell, taste, touch) Environment Provenance Quality Uniqueness	Certification (e.g. 'Fair Trade', Organic) Origin labelling Packaging Source	*Ben and Jerry's* (ice-cream) – Small town values and environmental commitment (Vermont, West Virginia, USA) *Molson* (Beer) – Clarity, purity and ice-cold temperature (Denver, Colorado, USA)

Source: Author's research drawing from Anholt (2002: 233), Phau and Prendergast (2000: 164), Pike (2009a), Roth and Romeo (1992: 4) and Thakor and Kohli (1996: 34–35).

retailer's fabricated associations to fictional places to convey quality and provenance in the brand names of 'Oakham' chicken and 'Lochmuir' salmon.

Geographical associations can be problematic in constructing meaning and value in specific spatial and temporal market contexts. Actors then try to deny, weaken or attempt deliberately to construct and project different geographical associations in brands and branding. Aspirant 'global' brands have emerged attempting to articulate a somehow 'space-less' or, echoing Relph (1976), 'place-less' identity and meaning free from any particular spatial attachments and ties. 'Global' identities are created to communicate modernity, reach and ubiquity in tapping international markets worldwide. The motion picture industry centred on Hollywood, California, for example, utilizes the 'blockbuster' franchise model. It deliberately avoids particular geographical referents in its films, except those with widespread and even global resonance, to reap the scale economies of global distribution and marketing while minimizing potential mis-readings in specific geographical markets (Hoad 2012). Conceiving of geographical associations enables scrutiny of such ostensibly 'global' brands. Google, Microsoft and Toyota retain – whether they want to or not – spatial connections and connotations, and try carefully to manage their 'global' attributes in adapting to particular geographical markets (Hollis 2010; Holt *et al.* 2004). Indeed, the 'global', seemingly ubiquitous, hyper-mobile and geographically limitless brands – such as Coca-Cola, McDonalds and Nike – are heavily imbued with geographically contextualized notions of Americanization, Imperialism and modernity (see, for example, Goldman and Papson 1998; Ritzer 1998). Further, each has been 'hybridized' and adapted to particular local market contexts by brand owners when commercially advantageous such as 'meta-commodity' Coca-Cola's experience in Trinidad (Miller 1998: 170). Such 'global' brands are neither 'space-less' or 'place-less' and cannot escape their geographical associations. Actors involved have difficulty shedding especially the brand's national images (Papadopoulos 1993). Their meaning and value is geographically differentiated and uneven as they are consumed differently in different places (Jackson 2004).

Distinguishing the different kinds, extents and natures of geographical associations explains how some are meaningful but not necessarily valuable in specific spatial and temporal market settings. Brand and branding actors exercise selectivity in constructing elements and their geographical associations in branded goods and services commodities. At one end, actors play up certain desirable and valued meanings such as the heritage, quality and reputation connoted by particular places – for example, Milanese design or Thai silks. At the other, actors mask meaningful but less commercially valuable or even damaging elements – for example poor quality and value automotive and electronics products linked to specific producers and particular places including Japan and South Korea in the 1970s and China in the 1980s (Willmott 2010). Diversity and variety in the different types, degrees and characters of geographical associations afford brand and branding actors a rich and pliable palette. Central to the origination argument, such assets and resources are conditioned by the particular histories and geographies of branded goods and services commodities and their spatial and temporal market experiences and reputations. Working the boundaries between material, perceived and fictional geographical associations, deliberately ambiguous and vague spatial connections can be commercially functional to actors involved in specific brands and their branding in particular market times and spaces. EAST clothing, for example, evokes a:

> generalised 'ethnic look' rather than a specific connection with India ... this influence is not always tied directly to India in terms of production ... the EAST brand is sustained through a range of discourses about fabric, design, and handwork and by an engagement with a generalised 'ethnic' aesthetic which may sometimes be inspired by India but is not uniquely grounded in the sub-continent. (Dwyer and Jackson 2003: 277)

The meaning and value of such malleable geographical associations afford the EAST brand owners a degree of agency and spatial flexibility in their production, circulation, consumption and regulation arrangements.

Distinguishing geographical associations in branded commodities demonstrates the worth of conceiving of tensions and accommodations in the relationships between territorial and relational conceptions of space and place (see, for example, Agnew 2002; Amin 2004; Bulkeley 2005; Pike 2007; Jackson 2004). Geographical associations constituting brands and branding geographies can be interpreted as relational *and* territorial, bounded *and* unbounded, fluid *and* fixed, territorializing *and* de-territorializing (Pike 2009a). Openness to the complexity and contingency of potentially contrary and/or overlapping tendencies and the ways in which they can be accommodated by the participant actors is helpful in researching and making sense of brand and branding geographies and their unstable and temporary nature.

Advancing beyond the constraints of marketing's *nationally* focused approach to '*Country* of Origin' (Bilkey and Ness 1982) to engage other geographical scales, geographical associations in brands and branding can be framed in territorial terms at different spatial levels (Pike 2009a). Actors try to construct geographical associations to delineated, even jurisdictional, spatial entities in establishing, representing and regulating the connections and connotations of particular branded goods and services commodities and their branding. Sociological accounts have demonstrated the role of geographical context and circumscribed geographical connections in imbuing potentially anonymous and mass produced commodities with identities 'by linking ... [them] ...to an identifiable (if often entirely fictional) producer or inventor or a particular physical place' (Arvidsson 2005: 244; see also Goldman and Papson 2006; Molotch 2002, 2005). As explained in Chapter 3, the critique of '*Country* of Origin' in marketing rests upon its limited and spatially fixed idea of the geographical associations and origins of specific brands framed solely at a nationally delimited scale. The national can be important in particular cases, but it is not the only territorial scale at which geographical associations are framed and articulated by the actors involved. National level connections and connotations are meaningful and valuable for specific branded goods and services commodities in certain market times and spaces. Actors managing Swatch watches, for example, have condensed '*S*wiss' and '*watch*' in its brand name. This demonstrates 'the way ... place of origin may be deliberately designed into the interface of the brand. This ... enables Swatch products to sell by securing the trust of (certain) consumers, providing a guarantee of quality, by

Table 2.7 Scales of geographical associations in brands and branding

Scale	*Examples*
Supra-national	European, Latin American
National	Brazilian, Japanese
Sub-national administrative	Bavarian, Californian
'National'	Catalan, Scottish
Pan-regional	Northern, Southern
Regional	North Eastern, South Western
Sub-regional or local	Bay Area, Downtown
Urban	Milanese, Parisian
Neighbourhood	Upper East Side, Knightsbridge
Street	Saville Row, Madison Avenue

tying the brand to an origin' (Lury 2004: 54). Space and place as territory can be part of the geographical associations of branded goods and services commodities and their branding. Spatial connections and connotations forged by producers, circulators, consumers and regulators draw upon and/or delimit territories at a range of spatial scales (Table 2.7). Examples of meaningful and valuable characteristics of brands situated and embedded in territorially defined spaces and places include Latin American Coffee, Swabian engineering ingenuity from Baden Württemburg, Catalan design, southern European cuisine, American western frontier spirit, Downtown Miami style, Parisian cool, Knightsbridge exclusivity and Saville Row quality.

Critically, geographical associations in brands and branding are not only scalar and territorial but relational and networked too. Unbounded, fluid and de-territorializing spaces and places that stretch through circuits and networks in and beyond clearly defined and delineated territories and spatial scales are evident in the geographical associations of brands and branding. Liz Moor (2007: 9) sees branding drawing upon more open and porous understandings of spatial entities because it seeks 'to meaningfully pattern units of information and link them across spaces'. Harvey Molotch (2002: 678) too claims that geographically associated brands support distanciated consumption because 'Through purchase, consumers in effect cannibalize a distant locale even without actually going there, taking in some of its social and cultural cachet'. In combining territorial and relational understandings, actors can utilize brands and branding as devices to create differentiated meaning and value. They borrow and attempt to transcend and/or hybridize, 'mash-up' and mix the material geographical associations of the territories of specific spaces and particular places. Nebahat Tokatli (2013) demonstrates how the modernization of the luxury fashion brand Gucci drew upon geographical associations based upon the look, style and sensibility of the City of Los Angeles, California. Even though the American designer Tom Ford was from Austin, Texas, in the United States and worked in Paris, France, the brand was headquartered in Florence, Italy, and the Gucci branded clothing was manufactured in Italy and, increasingly, internationally (see also Tokatli 2012b). Nigel Hollis (2010: 182) identifies the emergence of such 'brandspace' that is simultaneously relational and territorial because it 'isn't tied to any particular space or time; rather it is owned by the brand ... carving out an emotionally based territory for the brand communication'. In the digital era of real and virtual spaces, sociological accounts have gone furthest in developing a relational understanding of brands as interactional symbols, signs and logos. For Celia Lury (2004: 50), media pluralization has unleashed an unbounded spatiality for brands because 'The interface of the brand is not ... to be located in a single place, at a single time. Rather ... it is distributed across a number of surfaces (...products and packaging), screens (television, computers, cinemas) or sites (retail outlets, advertising hoardings...)' (see also Power and Hauge 2008; Power and Jansson 2011).

Dualistic conceptions that contrast *either* territorial *or* relational notions of space and place are poorly equipped to understand and explain the often complex, contingent and even contradictory ways in which the geographies of brands and branding unfold. Transformations and disintermediations through digitized means of communication and consumption of branded goods and services, for example, render the spaces and places of brands and branding more open, porous and relational. Real and/or virtual 'spaces of brands' (Hudson 2005: 68) are constructed within which otherwise disparate branded goods and services commodities are assembled. Each is inter-connected to create particular geographically associated experiences even without being materially and physically located in a specific place. Yet, even in on-line virtual space, 'e-branding' often needs adaptation and 'localization' in territorially demarcated and inflected ways because language, symbols, colours and consumer preferences remain heterogeneous and geographically differentiated (Ibeh *et al.* 2005). Indeed, Martin Kornberger (2010: 251) argues that 'The brand plays an even bigger role online than offline. ... Given that there are literally zero switching costs, the only thing that might keep people coming back to one particular site is an attractive brand.' Tensions and geographical differentiation are evident in the regulation of brand and branding geographies too. Efforts spatially to circumscribe ownership, copyright and other inescapably geographical associations of brands are always in flux because 'bounded jurisdictional spaces' in systems of governance and regulation are continually disrupted and shaped by shifting geographies of economy and value (Lee 2006: 418).

The forms, degrees and characteristics of geographical associations mobilized and articulated by actors through brands and branding unfold in 'unbounded', relational, *and* 'bounded', territorial, space and place over time in contingent ways. This more nuanced approach to the geographies of brands and branding makes Naomi Klein's (2000: xvii) 'post-national vision' of a more 'de-materialized', 'weightless' economy in which what 'companies produced primarily were not things ... but *images* of their brands' (2000: 4, original author's emphasis) appear somewhat overdone. Naomi Klein's (2000) *No Logo* has been vitally important in raising awareness of branded capitalism's political economy. The approach to brand and branding geographies articulated here moves forward by demonstrating how the geographical associations of branded objects and branding processes are more complex, diverse and variable, and benefit from explanation of their forms, extents and natures in the context of empirical research.

Despite the growing prominence of brands and branding, the explanations and politics of how their geographies are involved in any reconnecting (Hartwick 2000) or 'getting with' (Castree 2001: 1521) the commodity fetish remain underdeveloped. If 'an effect of branding is to distance the commodity from the social relations and conditions involved in its production by layering it with myths and symbols' (Edensor and Kothari 2006: 332), conceiving of the geographical entanglements of commodity brands and branding provides a way to analyse their uneven development and politics. This is an important but difficult task because:

> Branding in its current form takes this process [fetishization] a step further, promoting, in a sense, a fetishization of the fetish: that is, the commodification of the reified image of the commodity itself. Effectively, branding not only makes the 'mystical veil' which hides the social origins of the commodity that much thicker, but creates a veritable industry for the production and circulation of mystical veils, and devises methods for knitting these veils together to give the illusion of totality (Greenberg 2008: 31).

In the uneven geographical development of branded commodities and their branding, Noel Castree's (2001: 1520) 'important critical work' of de-fetishization remains unfinished. Lifting the mystical veils of brands and branding constructed by actors 'requires us to go beyond polarised debates about authenticity, "unveiling" the commodity fetish and revealing the exploitative social relations that are concealed beneath the commodity form' to instead articulate 'a more complex argument about the discursive process of appropriation ... to understand the cultural processes through which meanings are manufactured *as much as* the political-economic processes through which alienation and exploitation occur' (Jackson *et al.* 2007: 328–9, emphasis added). The notion of geographical associations contributes in two ways. First, it captures the material, discursive and symbolic in the diversity and variety revealed in cultural economy approaches (e.g. Cook and Harrison 2003). Second, it connects cultural construction with the systematizing rationales of contemporary capitalism – accumulation, competition, differentiation and innovation – emphasized but not reducible to any singular outcome by more culturally sensitive political economy (e.g. Castree 2001, 2004; Watts 2005).

Brands and branding in spatial circuits of meaning and value

Situating brands and branding in geographical context through conceptualizing geographical associations enables theorization of their role and importance in spatial circuits of meaning and value. Brands and branding are integral to the dialectic between spaces and circuits of value and meaning as well as representations of the 'economic' in markets (Hudson 2005; Sayer 2001). As Ray Hudson (2005: 68) explains, 'brand owners frequently present branded objects in themed spaces – parks, restaurants, pubs and shops – or contribute to the elaboration of themed lifestyles through the sponsoring of events and activities ... this creation of such "(hallucinatory) spaces of brands" exemplifies the dialectic between spaces and circuits of meaning'. Together with branding, 'the creation and promotion of brands are … pivotal in realising the surplus value embodied in commodities and so in helping assure the smooth flow of value and expansion of capital' (Hudson 2005: 76). The object of the brand and the process of branding provide means for the attempts of actors to create, cohere and stabilize the geographical associations constituting their meaning and value in spatial circuits. Dominic Power and Atle Hauge (2008: 125) interpret brands in this way as 'institutions' and branding as an 'institutional setting that lends structure to many industries' underlying economic arrangements and market processes' in which 'economic actors, not least firms, use brands as a core focus for their innovative and competitive activities and their market actions'. As Dominic Power and Atle Hauge (2008) acknowledge, there are limits to the fixity of such processes of 'institutionalization' through brands and branding. But critically important are the ways in which brands and branding are utilized in attempts to afford coherence to value and meaning in spatial circuits of capital. This is because 'particular objects need to maintain a degree of stability of meaning in order that they can perform as commodities and so enable markets to be (re)produced' (Hudson 2008: 430).

Brands and branding provide an object and a process for the efforts of actors to construct and fix what branded commodities mean and are worth in particular geographically situated market contexts and time periods. Their meaning and value are inextricably related to but not simply reducible to the brand's geographical associations. Any fixity and stability achieved by actors is only ever a temporary and partial accomplishment. The restless and disruptive rationales of accumulation, competition, differentiation, innovation, changing consumer tastes and so forth constantly threaten to recast and rupture the meaning and value in a brand and its branding in specific spatial and temporal market settings. As Hugh Willmott (2010: 526) acknowledges, 'there is often greater potential for generating surplus from intangibles like branding – even allowing for the vulnerability of brand equity to displacement by innovation, shifts in fashion or risk of its destruction by consumer activitists'. Brand and branding actors recognize this phenomenon:

> A brand must be managed and developed. It has to be constantly enriched and touched up to keep it in step with the times. Like a plant, it must be cultivated; it requires constant care if it is to stay alive, remain in fashion, and make a modern, likeable, attractive statement. … A brand's decline is an inevitable phenomenon if nothing is done to counteract it. The reasons, both internal and external, are numerous … errors in the brand's management – loss of relevance in a market, inefficient operations, incoherent strategies, inappropriate investments … competition … through the wide use of its products and the repeated broadcasting of its advertising, engenders a certain demystification. It loses some of its mystery, and thereby part of its attraction. In the contemporary context of the race toward novelty, this wearing effect of success is more rapid than ever … the evolution of the fundamental trends of our civilisation: needs, fashions, technology, tastes. (Chevalier and Mazzalovo 2004: 89, 128–9)

Logics and tendencies embedded in cognitive-cultural capitalism present on-going risks capable of undermining the coherence and stability of brands and branding as devices for continued value flow and capital expansion in spatial circuits. Meaning and value are not simply fixed or given by brand owners, stamped or marked on the finished goods and services commodities as they leave the factory or are communicated from the office. Because brands are economic, social, cultural, ecological and political they are inherently fluid, unstable, 'socially negotiated' (Power and Hauge 2008: 130) and 'fragile things' (Kornberger 2010: 53). Some are even contested over space and time through market rivalry and consumer activist dissent. When geographical associations constitute central elements of the meaning and value of brands, the on-going and evolving nature of space and place ensure that the fixing, cohering and stabilizing efforts of brand actors is only ever partial and temporary, even though brand meaning and value can in specific cases be enduring in particular times and spaces. The work of brand and branding actors is always unfinished and ongoing.

South Korean electronics group Samsung, for example, sought to rid itself of a reputation for low prices and poor quality geographically associated with new entrant producers from East Asia during the 1970s and 1980s. The brand owner sought proactively to reconstruct its geographical associations especially in profitable advanced country markets. Investments were focused on new product ranges, quality and brand building and promotion including high profile celebrity endorsement and sponsorship deals (Willmott 2010). Amidst the Asian financial crisis in 1996, Lee Kun-Hee, Samsung Chairman, aimed to move the company up the value chain from its Original Equipment Manufacturer (OEM) status by building its brand and R&D capabilities to 'raise brand value, which is a leading intangible asset and the source of corporate competitiveness, to the global level' (quoted in Samsung 2007: 1). Building a single strong 'masterbrand' was the strategy. Sub-brands including Plano, Tantus, Yepp and Wiseview were discontinued, and 17 new design centres were established in major cities around the world (Hollis 2010). Recent flagship initiative the 'Samsung Experience' in New York uses a showcase store located in a prime retail site in a trend-setting global city and experiential marketing techniques to expose consumers to new products prior to general release. The intention is to implicate the new branded commodities into consumer lifestyles as a means to inspire loyalty, stimulate word of mouth circulation and encourage repeat purchase and consumption (Moor 2007). Yet, such attempts to create, cohere and stabilize meaning and value are not the same always and everywhere because of the geographical differentiation inscribed in brands and branding. They can mean and be valued differently by different people in different places. Samsung dominated its domestic market but initially downplayed its particular national geographical associations internationally until it was able to articulate and substantiate a reputation for higher quality (Quelch and Jocz 2012). Stereotypical images and reputations of particular countries are often entrenched and endure, affecting how products and services associated with those countries are perceived, and sometimes resisting reshaping efforts such as advertising, national export drives and other promotions (Phau and Prendergast 1999).

Although a historically longstanding and enduring brand, Pear's Soap gradually lost its meaning and value in its specific spatial and temporal market situations. Hairdresser Andrew Pears patented Pear's Soap in London in 1789 (Haig 2004b). The transparent design provided a distinctive and differentiated feature of the brand. In Britain during the Victorian era, Pear's Soap was in the vanguard of the early advertising efforts for mass produced and packaged goods, creating and circulating a strong brand name and identity. Endorsements were utilized to raise brand awareness. These included Sir Erasmus Wilson, President of the Royal College of Surgeons, claiming that Pear's Soap had 'the properties of an efficient yet mild detergent without any of the objectionable properties of ordinary soaps' (quoted in Haig 2004b: 219). Entry to the American market was secured through national level advertising and the endorsement of religious leaders 'to equate cleanliness, and Pear's particularly, with Godliness' (Haig 2004b: 220). The Pear's brand gained commercial success in the United Kingdom and America, becoming 'part of everyday life on both sides of the Atlantic' (Haig 2004b: 220). As incomes and prosperity rose in the advanced economies, soap was at the forefront of consumer culture. It supported the growth of consumer goods companies Procter and Gamble and Unilever, and was used to pioneer market research, TV advertising and sponsorship of 'Soap Opera' television series (Bunting 2001). Pear's maintained its sales momentum in an expanding market through the post-war period until the mid-1990s. By the late 1990s, the spatial and temporal circumstances of the market in the UK shifted adversely for the commercial fortunes of the Pear's brand. The market fragmented and undermined demand for the 'mass produced block' (Haig 2004b: 220). New liquid versions of shower gels, body washes and liquid soap dispensers from existing and new producers grew dramatically in a variety of new and existing brands, colours and fragrances. Blocks of soap expanded in a low volume craft segment that actors positioned as artisanal and authentic in contrast to the mass-produced brands. Pear's market share collapsed to 3% and marketing investment was reduced to zero for what was considered a dying brand. While Pear's is still sold in some markets such as India, owner Unilever discontinued the brand in the UK in 2000 as part of its strategic focus upon 400 'power brands' and elimination of 1200 weaker and less successful brands (Haig 2004b). The coherence of the brand in the UK unraveled as shifts in consumer tastes and technologies combined with declining marketing and advertising to undermine its meaning and value in its market times and spaces. Contrasting Pear's demise with the continued marketing support and commercial success of Unilever's Dove brand soap bar, Matt Haig (2004b: 221) concluded 'Pear's was a brand built on advertising and when that advertising support was gradually taken away, the brand identity became irrelevant'.

In spatial circuits, the creation of value is 'manufacturing meaning' (Jackson *et al.* 2011: 59) too as 'commodity production itself becomes ever more deeply infused with aesthetic and semiotic meaning' (Scott 2007: 1474). The growing importance of intangible assets demonstrates this trend. Value creation shifted from direct labour power to tangible assets such as machinery and factories as part of the substitution of labour for capital during the nineteenth century industrial age and later era of mass manufacture (Mudambi 2008). More recently, the source of value creation has moved towards intangible assets, including R&D, intellectual property, patents, copyrights and the goodwill incorporated in brands and branding activities (Lury and Moor 2010) (Figure 2.3). The proportion of market valuation of the Standard and Poor Top 500 American firms made up of intangible assets, for example, rose from 38% in 1982 to 85% in 2001. By 2004, investment in intangibles in the American economy was estimated at over 8% of GDP or $1 trillion – 33% was software, 33% intellectual

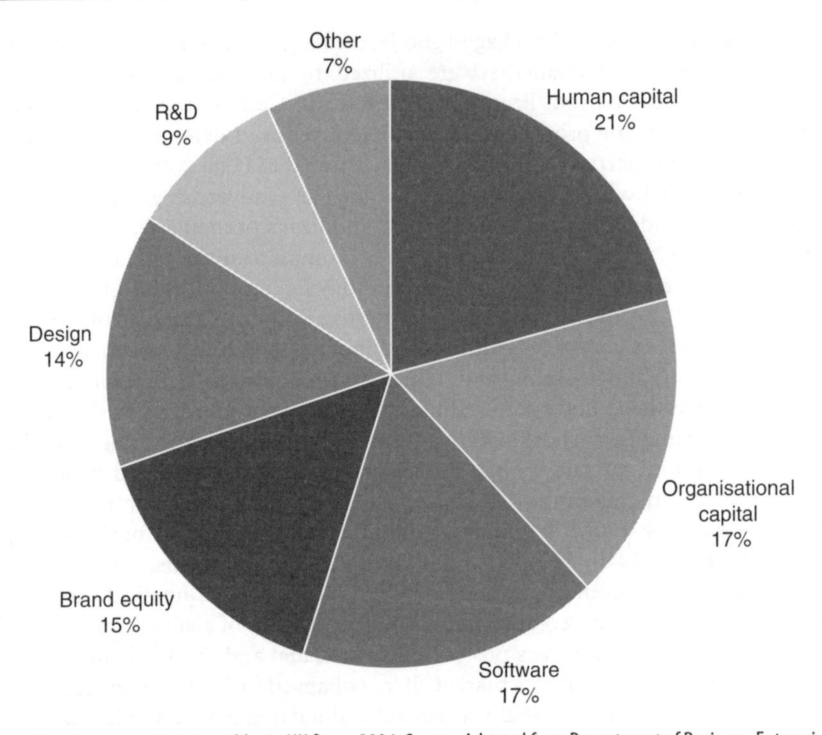

Figure 2.3 Investment in intangibles in UK firms, 2004. Source: Adapted from Department of Business, Enterprise and Regulatory Reform (DBERR) (2008) data.

property including patents and copyright and 33% advertising and marketing including branding (Nakamura 2003).

Brands and branding do not simply appear fully formed on branded goods or services commodities at the point of exchange and sale. Brands and branding have permeated moments *within* and *across* spatial circuits in meaningful and valuable ways. Taking the basic Marxian circuit of capital (M–C … P … C'–M'), the object of the brand and process of branding shape and, in turn, are shaped by the spatial connections and connotations throughout the full spatial circuit. The financialized meaning and value of geographically associated brands condition the access to capital in money form for actors even to initiate an accumulation circuit (Lindemann 2010; Willmott 2010). When money capital is invested to purchase the initial commodities to assemble the means of production, labour power and raw materials (whether from nature or intermediate goods and services suppliers), in cognitive-cultural capitalism such commodities can be branded and imbued with geographical associations. Means of production might be Beohringer machine tools from Germany in a factory or Reuters news services from New York on a financial trading floor. Who and where provides these branded inputs can matter to the meaning and value of the subsequent branded goods produced and/or services delivered. Labour power may be provided by people in places renowned for their combinations of skills, productivity and wages and able to articulate a reputable brand in the international trade in tasks. This is especially the case in non-routine activities requiring high levels of interpersonal interaction (Kemeny and Rigby 2012). Examples include *haute couture* fashion designers in Paris (Scott 2000) or emerging market investment analysts in Bangalore (Dossani and Kenney 2007). Raw materials especially embody meaning and value in their provenance and its branded articulation, for example gourmet quality saffron spice from La Mancha (Chapter 7) or particular teak wood varieties from Thailand (Bryant 2012).

In the labour process, meaning and value creation go hand in hand because of who is involved and where it takes place (Sennett 2006). For branded goods and services commodities, this might constitute a handcrafted timepiece produced by a watch-maker in Geneva (Glasmeier 2000) or services delivered from a call centre in Glasgow for a UK-based bank (Taylor *et al.* 2002). Inescapably geographical brands and branding are integral to the intertwining of meaning and value in the differentiation of output for sale in particular socio-spatial market contexts:

> To be sold, [*branded*] commodities must be seen to be useful to their purchasers – to have use value and this in turn implies that they are seen as meaningful in the context of their life worlds. Such meanings may relate to strictly utilitarian aspects of commodities (for example, sheet steel purchased by automobile producers) or, increasingly in the case of final consumers, their affective dimensions and culturally coded symbolic meanings. (Hudson 2008: 429; emphasis added)

Meaning and value are created in this moment in diverse and varied geographical settings. In Nike-branded trainers purchased in Niketown, Shanghai (Goldman and Papson 1998), for example, or O_2 mobile telephony bought through roaming services in São Paulo. Concluding a loop of the full circuit, brands and branding are central in the sale and purchase of commodities in consumption. Surplus value is realized at this moment as expanded money capital for extraction or re-investment in further circuits (Hudson 2008). Geographical associations are circulated and exchanged in brands through the processes of distribution, marketing and sale by what Vance Packard (1980: 171) termed the 'symbol manipulators' as 'such spaces are simultaneously material sites for commodity exchange and symbolic and metaphorical territories' (Hudson 2005: 145). Consumers interpret and make sense of the claims and representations of brands and branding, framing and shaping their agency. Branded retail outlets, for example, are located to create meaning and value for the participant actors from their particular situation in specific shopping districts in certain places such as 5th Avenue, New York City, or the Ginza, Tokyo.

A common focus in social science considers the role of brands and branding in the consumption moment for final consumer goods and services commodities. Theorizing brands and branding as integral to meaning and value construction *throughout* the spatial circuits of cognitive-cultural capitalism advances our understanding. It provides a way to grasp how the geographical in brands and branding unfolds and – critically for the aim here of conceptualizing and theorizing origination – how, why, where and with whom such agency resides. Despite recognition of the importance of spatial circuits, recent

Table 2.8 Brand and branding actors

Actor	*Examples*
Producers	Brand owners, designers, manufacturers, 'place-makers', residents
Circulators	Advertisers, bloggers, journalists, marketers, media
Consumers	Shoppers, residents, tourists, users, visitors
Regulators	Government departments, trademark authorities, local councils, export agencies, intellectual property advisers, business associations

Source: Author's research.

research has privileged sites of production, consumption and, to a lesser degree, circulation as particularly significant and fluid moments in which value and meaning intersect through brands and branding. Sociological views, in particular, emphasize the centrality of the consumption moment. Brands are seen to act as trust mechanisms that generate social dependencies by providing consumers with 'real informational, interactional and symbolic benefits' (Holt 2006a: 300) from which brand owners extract economic rents. Echoing Colin Campbell's (2005) craft consumer, Adam Arvidsson (2005: 237, 244) too claims that 'the meaning-making activity of consumers … forms the basis of brand value' and branding establishes and shapes the 'context of consumption'.

Consumer agency is certainly integral to the meaning and value of brands and branding. It is especially critical in contesting and reworking the intentions of brand owners, for example through the role of local consumption cultures in adapting 'global' brands in local markets such as Cadburys in China (Jackson 2004). But the recent focus upon novel if relatively small scale forms of producer–consumer relationships and their importance for innovation (see, for example, Arvidsson 2006; Thrift 2006) present worthwhile but only partial accounts. As Martin Kornberger (2010: 43) puts it 'brand meaning cannot be determined by any one agent, be it a manager or consumer'. Such work can benefit from more thoroughgoing integration within the broader spatial circuits of value *and* meaning that connect production, circulation, consumption and their regulation. In elaborating a more encompassing conceptual framework, the varied actors involved in brands and branding are better conceived as the producers, circulators, consumers and regulators of meaning and value interrelated within and across spatial circuits (Table 2.8). This approach enables the identification of the actors involved and assessment of their roles, positions and claims within the broader spatial circuit within which they are related and animated. It facilitates engagement with actors from the public, private and civic spheres as well as hybrid and innovative forms of social and institutional organizations.

Focusing upon brands and branding within spatial circuits advances our understanding of their geographies because their utilization in meaning and value creation extends *beyond* the point of production. As John Allen (2002: 41) argued:

> the mix of images in advertising, the sign value of material objects, the semiotic work of branding … this symbolic activity adds up to an aestheticization of the economic, *which takes place within the sphere of production as well as in the circuits of exchange and consumption* [emphasis added].

The process of meaning and value making unfolds in 'the sphere of circulation where the labour of unwaged user-consumers is seen to participate in providing content, and thereby building brand equity' (Willmott 2010: 518). 'Co-production' of meaning and value is evident next in consumption. Consumers identify with and attribute qualities to brands, mobilizing, articulating and sharing meaning and value within spatially extensive and especially recently on-line 'brand communities' (Arvidsson 2005).

Meaning and value are created too in the regulation of production, circulation and consumption. Relatively neglected in recent accounts, the regulation of brands and branding requires more in-depth consideration within and through production (e.g. IT, trademarks, quality accreditation, geographical indications of origin), circulation (e.g. advertising standards, copyright restrictions, ethical codes) and consumption (e.g. rules of origin, product labelling, recycling requirements) (Lury 2004; Morgan *et al.* 2006). Actors use regulatory devices in attempts to construct and fix meaning and value in certain market situations. Such practices aim to control, institutionalize and appropriate values and meanings geographically associated in branded goods and services and branding practices. These devices are 'key components' of a brand and comprise 'brands names, logos, packaging designs, color schemes, shapes and smells' as well as 'patents for technologies and formulas' (Lindemann 2010: 7). Regulation pervades the worlds of real and virtual spaces. The focus on regulation is critical too in emphasizing the political and economic importance of legal and financial ownership and control of brands and branding by specific actors with particular interests. This reading is needed to temper some views of brands that emphasize the co-production of meaning and value where consumers are seen as at least part or in some way equal 'owners' of brands (see, for example, Kornberger 2010). Regulation has become more important recently with the growth of counterfeit goods and services and the inability of consumers to identify fakes especially when the quality is high (Phau and Prendergast 1999).

The regulation of the geographical associations of origin for manufactured products are governed by rules of origin. These regulations form part of the trade rules between nation states agreed and negotiated through inter-governmental structures by the World Trade Organization (WTO). Rules of origin set out criteria that define where specific products are made. The rules are integral to trade policy instruments and discrimination between countries in their trading relationships. The rules help in addressing issues including anti-dumping, countervailing duties (charged to counter export subsidies), the marking of origin and 'made in…' labels for products, preferential tariffs, quotas and safeguard measures as well as providing the basis for trade statistics compilation. For countries, rules of origin are potential non-tariff barriers to trade, raising costs for producers from particular origins and enabling other producers to segment markets (Krishna 2005). But, as the World Trade Organization (2013: 1) puts it, 'made in … where?' It is increasingly difficult in the context of complex and internationalized and even global value chains precisely to identify origin: 'determining where a product comes from is no longer easy when raw materials and parts criss-cross the globe to be used as inputs in scattered manufacturing plants' making regulating 'rules of origin' 'complicated by globalization and the way a product can be processed in several countries before it is ready for the market' (WTO 2013: 1). Emerging issues have further complicated the task. These concerns include the growing number of preferential trading relationships between countries with bespoke rules of origin, increases in origin disputes especially in relation to quota agreements for example in the Multi-Fibre Agreement in textiles and clothing and in steel trade, and increased bilateral anti-dumping disputes where attempts have been made to circumvent duties by the use of facilities located in third countries.

Current regulating principles have been developed through incremental intergovernmental agreement and attempts at harmonization and avoidance of barriers to trade in the post-war period, first in the General Agreement on Tariffs and Trade (GATT) and then the WTO. The principles aims to ensure member nation states make their individual 'rules of origin … transparent' so 'that they do not have restricting, distorting or disruptive effects on international trade; that they are administered in a consistent, uniform, impartial and reasonable manner; and that they are based on a positive standard (in other words, they should state what *does confer origin rather than what does not*)' (WTO 2013: 1). Within the EU, the rules of origin define origin as 'the "economic" nationality of goods in international trade'. The rules distinguish between 'non-preferential' to determine the origin of products and 'preferential' to identify specific goods from particular countries that can benefit from beneficial arrangements such as reduced or zero duties (European Commission 2013: 1). Critical to the EU rules are whether or not the goods are procured or produced from a single or multiple countries, complicated by the origin of materials, where the final and/or most substantive stage of production activities is located, and the relative value-added contributed to the final goods from activities based in each country. All of which issues are difficult to specify and measure (WTO 2013).

Rules of origin frame the regulations on product labelling that shape how geographical associations are recognized and articulated on specific branded products. Product labelling standards set by international bodies interact with different national institutions and their jurisdictional territories to set requirements on what brand owners legally have to disclose in their products and services. Such regulations include, for example, where a branded good was assembled or produced. The aspiration with product labelling regulations is to provide as near to 'perfect' information for buyers and sellers in markets to enable them to make informed choices, reducing the risk of market failure through asymmetric information and protecting consumer interests. Supra-national and national regulations interact to produce particular requirements. These are based upon issues of packaging and labelling compliance, what the product is made of and how it is made, translating labelling into local languages and marking the origin of the goods to the level of place and date of manufacture. Forms of regulation seek to control geographical associations in branded goods and services commodities too through frameworks such as geographical indications (GIs). GIs protect spatial references to brand provenance and attribute characteristics to geographical origin through marks that 'can be seen as attempts to tie particular qualities inherent in the *product* to particular qualities inherent in the *context of production*' (Parrott *et al.* 2002: 246; original authors' emphasis) (See Chapter 7).

Critical in explaining the role of different actors – producers, circulators, consumers and regulators – in spatial circuits of brands and branding is recognizing their different interests and agency. As Michel Chevalier and Gérald Mazzalovo (2004: 178) put it:

> The managers of the different functions involved have different training and different specialities. The merchandiser wants to sell now; however, the identity manager cultivates the brand for the long term. The production manager wants to reduce costs; the brand identity manager always looks to improve quality.

Such interests can converge and diverge, even conflict, to varying degrees in the spatial circuits of branded commodities. Constructing and fixing meaning and value through geographical associations in specific market situations is challenging for the actors involved.

Summary and conclusions

This chapter established the geographical reading of brands and branding, addressing their relatively under-researched position and laying out the conceptual and theoretical foundations of origination. The object of the brand was defined as an identifiable kind of good and/or service commodity comprising differentiated characteristics. In the framework of 'brand equity', such attributes constitute tangible and intangible assets or liabilities enhancing or denuding the meaning and value in brands and branding for the actors involved. Branding was explained as the process of making meaning from the articulation and communication of the characteristic and distinctive elements of the brand in differentiated ways that add value to goods and services commodities in competitive spatial and temporal market settings. Together brand and branding connect value *and* meaning as integral elements of contemporary 'cognitive-cultural capitalism' (Scott 2007: 1466). Dispelling the possibility of brands and branding as 'spaceless concepts' (Lee 2002: 334), the ways in which brands and branding are geographical was set out. First, it was demonstrated how *inescapable* spatial connections and connotations are integral and longstanding – but not always the only – categories of brand and branding definition, value and meaning. Second, the geographically differentiated manifestation and circulation of brands and branding across space and time was illuminated. Last, the interrelations between brands and branding and geographically uneven development were explained through accumulation, competition, differentiation and innovation rationales orchestrating and reflecting socio-spatial inequality and hierarchy. The conceptualization and theorization of how, why, by whom, where and in what ways geographies are entwined in branded goods and services commodities and their branding was then explained. Drawing from the debate about entanglement, geographical associations were introduced and explained as the characteristic elements of the identifiable branded commodity and branding process that connect and/or connote particular 'geographical imaginaries' (Jackson 2002: 3). Distinctions were explained in the kinds (e.g. material, symbolic, discursive, visual and aural), extents (e.g. strong, weak) and natures (e.g. inherent, authentic, fictitious, hybrid) of geographical associations. The expression of geographical associations was explained in the interplay, tensions and accommodations between territorial scales *and* relational networks.

The theorization of brand and branding geographies was explained in the creation, circulation and valorization of differentiated meaning *and* value in branded goods and services by actors: producers, circulators, consumers and regulators. Emphasizing their interrelation through spatial circuits including *and* beyond the point of production, these actors were interpreted as utilizing brands and branding as a means selectively to construct, cohere and stabilize geographical associations in branded goods and services in commercially valuable and meaningful ways in specific market times and spaces. Emphasizing and/or obscuring spatial connections and connotations, actors try to secure a degree of fixity for brands to perform as commodities to enable accumulation but could achieve this only on a temporary and partial basis. The rationales of accumulation, competition, differentiation and innovation propelled by and reflective of brand and branding actors disrupt the meaning and value of certain geographical associations in specific brands and their branding in particular periods and places. Disruptive logics unsettle the coherence and stability of brands as devices for ensuring the continuous flow of value and expansion of capital in spatial circuits.

Chapter Three
Origination

Introduction

Origination seeks, first, to understand and explain exactly what actors do in efforts to construct geographical associations in branded commodities and their branding to imbue and capture meaning and value for specific goods and services commodities in certain spatial and temporal market settings. And, second, to interpret how actors in spatial circuits enact and respond to accumulation, competition, differentiation and innovation logics that disrupt the temporary fixes of geographical associations integral to the meaning and value of their brands and branding. The issues of geographical origin(s) and provenance, the questioning of and transitions beyond '*Country* of Origin', and the socio-spatial histories of branded commodities and their branding are addressed. As the conceptual and theoretical core of the book, origination is then introduced, explained and exemplified. The methodological and analytical challenges of researching origination in the geographies of brands and branding are then tackled.

Geographical origin(s) and provenance

The geographical origin(s) of goods and services commodities are historically longstanding and enduring characteristics. Provenance is defined literally as place of origin. Depending upon the specific commodity and the spatial and temporal market situation, provenance can be a valuable and meaningful marker of particular attributes such as authenticity, quality and reliability. As established in Chapter 2, goods and services

Origination: The Geographies of Brands and Branding, First Edition. Andy Pike.
© 2015 John Wiley & Sons, Ltd. Published 2015 by John Wiley & Sons, Ltd.

commodities are inescapably imbued with geographical associations. This is because, as Harvey Molotch (2002: 665, 686) puts it, they 'contain – in the details of their fabrication and outcome – the places of their origin. ... Place gets into goods by the way its elements managed to combine and the stuff shows it.' Søren Askegaard (2006: 94) too sees that 'there is behind each brand a built-in cultural reference that refers to an origin, even for "global" brands. But this place of origin will co-exist in each consumer's life with a large number of other potential spatial references.' The connection between place and brand is important. It is powerful in the creation of branded goods and services since 'Once the buyer is able to identify the provenance of a commodity, it is in the producer's interest to protect the integrity of his products against deceptive manipulation by intermediary traders. Brands are therefore always associated with the pre-packaging and sealing of products' (Fanselow 1990: 253). Where goods and services actually come from or are associated with – or crucially are *perceived* to come from or are associated with – are potentially important assets and liabilities in the framework of brand equity. Provenance provides facets and cues that actors attempt to render meaningful and valuable in the times and spaces of particular markets.

Historically, marks identifying and seeking to differentiate brands typically deployed geographical identifiers of especially particular places. Devices including crests, emblems, hallmarks and images of distinctive architecture, art, crafts, design, folklore, landmarks, people and textiles as well as producer and/or ingredient origins were used to communicate where a good or service was from and/or associated with (Fleming and Roth 1991; Room 1998). Geographical separation of producers and consumers underpins Douglas Holt's (2006a: 299) claim that brands have been 'elemental to markets since traders first marked their goods as a guarantee for customers who lived beyond face-to-face contact'. During the early period of industrialization, some brand owners tried to lose their geographical ties as part of connecting to wider markets. 'The Great Atlantic and Pacific Tea Company', for example, became A&P (Chevalier and Mazzolovo 2004). As the era of industrialization evolved, specific 'products-as-brands' making explicit geographical connections emerged from the nineteenth century. Producers sought differentiation and distinction through specific associations with particular places within national market contexts: 'Each major nation had its own steelworks, munitions factories, chemical and soap plants, shipyards and so on, with individual idiosyncrasies. Everything from bread and cakes to architectural styles and modes of dress had a national flavour, and they were often very important' (Olins 2003: 136). Examples that endure involving spatial references to the national and other scales include Brazilian coffee, Cadbury's Bournville chocolate, Hollywood films, Indian tea, Japanese electronics, London insurance, Parisian fashion clothing and Sheffield Steel (Moor 2007; Papadopoulos and Heslop 1993).

As elements of regulatory requirements and packaging, 'Made in...' labels have been used to identify the geographical origins of branded goods for over a century (Morello 1984). Examples are manifold, typically situated at the national level as a result of the resonant meaning and value of '*Country* of Origin'. The enduring meaning and value of 'Made in Germany', for example, reflects its historical engineering capability and reputation as a manufacturing nation of high quality goods, despite globalization and European integration pressures (Harding and Paterson 2000).

Figure 3.1 'Hella – Quality made in Germany'. Source: Hella KGaA Hueck & Co.

As Wally Olins (2003: 137) puts it, 'the halo effect of Mercedes, Siemens and brands like them still continues to give German engineering products credibility' (Figure 3.1). National 'Made in…' claims have extended geographically in the wake of the globalizing division of labour to incorporate new emergent nation states such as Bosnia-Hercegovina, the Czech Republic and Slovakia. Awareness of *country* origin and image for product brands grew as part of internationalization from the 1950s. It was reinforced in the late 1970s and early 1980s by rules of origin regulation linked to international trade liberalization and national modernization and (re)industrialization strategies (e.g. national level 'local content' requirements to encourage and support 'local' productive capacity, infant industries and technological upgrading), economic nationalism (e.g. 'Buy American' and 'Buy British' campaigns to stimulate domestic demand, support domestic producers and reduce import penetration and trade imbalances), and increasing and discriminating use of origin identifications as part of creating differentiated value and meaning in competitive market contexts (Papadopoulos and Heslop 1993).

Since the early 1990s, influential branding commentators such as Nicolas Papadopoulos (1993: 10) have noted the growth in 'origin identifiers' as part of product marketing. Reflecting Michel Callon's (2002) 'economy of qualities' and what Igor Kopytoff (1986) calls the imperative to 'singularize' commodities, the geographical associations of a singular *origin* or plural *origins* have grown in importance and visibility. They have become potentially durable and defensible sources of value and meaning underpinning differentiation. 'Country image identifiers' (Papadopolous 1993: 17) and product 'nationalities' (Phau and Prendergast 2000: 164) have been seen as increasingly important because the logics of competition and standardization in globalizing markets threaten uniqueness and stimulate demands for authenticity (Beverland 2009; Storper 1995). Provenance – place of origin – can in certain market settings provide branded goods and services commodities with valuable and meaningful attributes such as authenticity, quality and reliability.

In a rational choice framework, goods and services are conceived as presenting different types of cues that influence consumer agency. These comprise intrinsic cues relating to the characteristics of the good and/or service and extrinsic cues constituting other related considerations (Bilkey and Nes 1982). Multiple intrinsic (e.g. design, fit, performance, taste) and extrinsic (e.g. brand, price, reputation, warranties) cues and facets together shape consumer perceptions and behaviour in relation to goods and services commodities (de Chernatony 2010). Longstanding research in marketing reveals how product-*country* image and origin – understood as an extrinsic cue – are often decisive in consumer decision-making. As explained in Chapter 2, such cues underpin the longstanding '*Country* of Origin' or 'Made in…' effect evident in consumer views of the geographically differentiated capabilities and historical reputations of countries for particular goods and services (Bass and Wilkie 1973; Bilkey and Nes 1982; Johansson 1993; Thakor and Kohli 1996). This work has traditionally remained focused on the national level. Indeed, national stereotypes infuse perceptions of specific commodities. For example, Peter van Ham's (2001: 2) caricature states that 'We all know that 'America' and 'Made in the U.S.A.' stand for individual freedom and prosperity; Hermès scarves and Beaujolais Nouveau evoke the French *art de vivre*; BMWs and Mercedes-Benzes drive with German efficiency and reliability'.

Understood as geographical associations that vary over time in their kind, extent and nature, origin cues can be utilized by branding actors to articulate and shape the assets and liabilities of brand equity that connote geographical origin or origins. A range of strategies, frameworks, techniques and practices have been deployed by branding actors for specific brands in particular market contexts over space and time. Nationally framed geographical associations have been prominent in the context of 'Country of Origin'. Examples include direct (e.g. British Airways, Nippon Steel) or indirect (e.g. national language use such as Gucci and Lamborghini) brand name reference, labelling (e.g. 'Made in…') and/or national framed geographical symbols in brand logos (e.g. national flag, emblem) (Riezebos 2003; Thakor and Kohli 1996).

Extending the focus to 'country-of-origin of *brand*' (Phau and Prendergast 2000: 159), multi-faceted product-country images work as either 'halo constructs' that shape consumer evaluations of products in conditions of complex and imperfect knowledge or 'summary constructs' that project product origin knowledge onto countries (Han 1989; Papadopoulos and Heslop 1993). Martin Roth and Jean Romeo (1992) identify a number of dimensions of country image that affect the evaluation of branded goods and services. These include efficiency, design, innovativeness, design, prestige, quality, reputation, workmanship and value (see Table 2.8). Such geographically rooted and/or inflected characteristics are utilized by actors to form the basis of geographical associations in attempts to construct and cohere meaning and value in branded commodities in spatial circuits. They enable 'producers to position their brands simply, strongly, and quickly' (Morello 1993: 288) and are 'used by consumers to reinforce, create, and bias initial perceptions of products' (Johansson 1993: 78). Yet, as discussed in Chapter 2, the geographical associations constructed by actors in brands and branding encompass and extend beyond the singular national frame of 'Country of Origin'.

Beyond '*Country* of Origin'

The internationalizing spatial division of labour and emerging geographical patterns of economic activities across the world have complicated and questioned understandings of the '*Country* of Origin' of brands and their branding. This disturbance has occurred for a number of related reasons. First, for increasingly complex manufactures with multiple component parts and sub-systems, it is no longer straightforward to assign a single and definitive national geographical origin to the finished branded good provided for sale in a particular spatial and temporal market setting (Dicken 2011). As Andreas Maurer (2013: 13) of the WTO notes: 'Industrial supply chains add to the "blurriness" of the country of origin concept as the value of an imported good does not necessarily fully originate from the geographical origin mentioned in custom documents'. Ian Phau and Gerard Prendergast (1999: 72) too claim that: 'The controversy of the relevance of country of origin continues to cloud the ambiguous tags on products with multi-country affiliation'. As forms of economic organization over space and time – whether conceived as global production networks (Coe *et al.* 2004, 2008; Henderson *et al.* 2002) or global value chains (Gereffi *et al.* 2005; Gibbon and Ponte 2008; Neilsen and Pritchard 2011) – become more complex, sophisticated and extend geographically, specific branded commodities can be composed of multiple parts and sub-assemblies and combine activities from myriad goods *and* services suppliers located across the globe. Ram Mudambi (2008: 704) demonstrates that 'technological advances, especially in the areas of information and communication technologies, have made it possible to disaggregate the firm's business processes into progressively finer slices'. For the integration and coordination of this finer grained international division of labour, two divergent strategies for control and location are evident:

> A vertical integration strategy emphasises taking advantage of 'linkage economies' whereby controlling multiple value chain activities enhances the efficiency and effectiveness of each one of them. In contrast, a specialization strategy focuses on identifying and controlling the creative heart of the value chain, while outsourcing all other activities … along the location dimension, a common pattern of geographical dispersion appears to be developing. Firms are increasingly implementing strategies to take advantage of the comparative advantages of locations. This results in a wider geographic dispersion of firms' activities. (Mudambi 2008: 702)

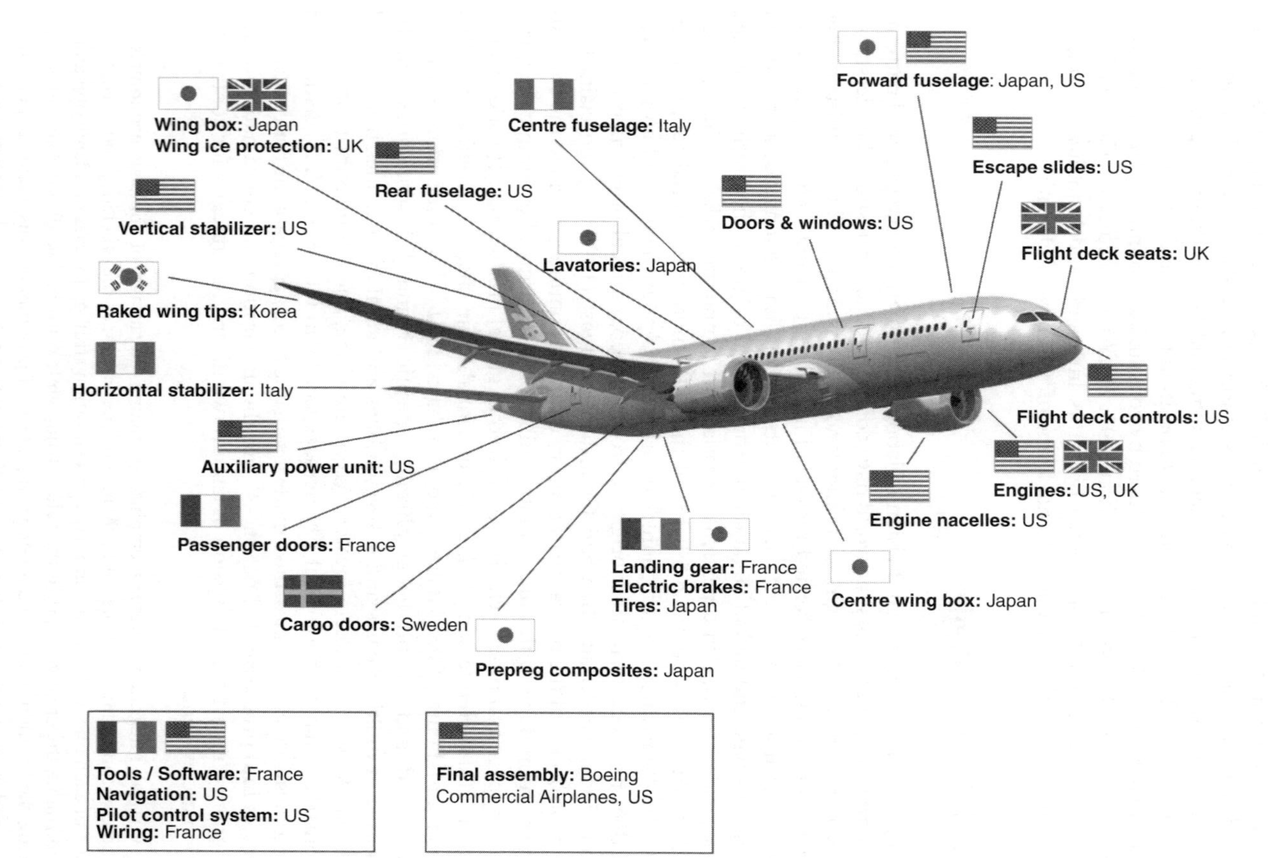

Figure 3.2 Global value chain for the Boeing 787 Dreamliner. Source: Sebastian Miroudot, World Bank.

Branded goods now represent 'hybrid configurations' for which there is no easily discernable single or multiple country cue. This is because of the:

> over-whelming increase in products that are manufactured in diverse locations around the world and marketed under a single brand name. Under such circumstances, country of origin is often viewed as a multidimensional construct involving a hybrid of factors that makes the distinction between the country of manufacture or assembly and the country of the company's home office. (Phau and Prendergast 1999: 1981)

Figure 3.2 illustrates Boeing's coordination and integration of a complex set of component parts, systems and activities to deliver a specific branded good and services in the particular internationalized value chain and spatial circuit for the 787 Dreamliner aircraft. Consumer goods and services too have become organized and coordinated in such complex and geographically expansive systems (Dicken 2011). Indeed, constituent goods and services commodities can be branded too, especially where they relate to activities central to the meaning and value of the final branded commodity. Many different varieties of branded laptop computers, for example, are marked with the additional 'Intel Inside' brand label to signify the micro-processor is provided by US micro-electronics company Intel and assure consumers of its quality, performance and reliability.

During the 1980s, such issues of geographical origin animated debates about the value and impact of investments in 'transplant' car assembly factories in America by Japanese producers such as Honda and Toyota (Mair *et al.* 1988). American consumer groups and the United Auto Workers union promoted 'Buy American' campaigns in support of the output and employment of the unionized factories of the 'Big 3' American-owned giants of Ford, General Motors and Chrysler. Others argued that the Honda and Toyota branded vehicles were also 'Made in the USA' by Japanese-owned producers in new often non-union transplants created through foreign direct investment. At the time, it was considered increasingly complicated to answer Robert Reich's (1990) question of 'Who is us?' The globalizing division of labour has proceeded apace since the 1980s, further complicating claims to any 'Made in…' origin. In some views, as consumer awareness and familiarity with products grows 'the use of the made-in label begins to diminish' (Phau and Prendergast 1999: 1981). As more countries develop their production capacity and skills, Country of Origin is claimed to be of relatively little importance (Sheth 1998). Examining the rise and popularity of 'global brands', Douglas Holt *et al.* (2004: 71) claimed that:

> Until recently, people's perceptions about quality for value and technological prowess were tied to the nations from which products originated. 'Made in the USA' was once important; so were Japanese quality and Italian design in some industries. Increasingly, however, a company's global stature indicates whether it excels on quality. We included measures for country-of-origin associations in our study as a basis for comparison and found that, while they are still important, they are only one-third as strong as the perceptions driven by a brand's 'globalness'.

The origin category has been opened up, extended and articulated by actors more closely to reflect – as well as in some cases obscure – which activities are undertaken, by whom and where. Driven by brand owners' differentiation strategies and regulatory standards, new forms of communication and labelling formats are emerging. These practices are grappling with the notion of multiple and connected – even 'global' – origins for particular production and service activities constituent of specific brands. These new markers of origin(s) are beginning to reference and represent the places of different functions including assembly, design, delivery, engineering, component sourcing and manufacture (Phau and Prendergast 2000).

Second, the 'Country of Origin' of brand has been complicated by the rise of services. In the continued transition to service-dominated economies, tradeable services have risen in importance facilitated by advances in information and communication technologies and the blurring between manufacturing and services because of the rising importance of intangibles (Dicken 2011; Mudambi 2008). The relational nature of services has traditionally troubled considerations of their provenance and origin. When face-to-face interactions dominated – for example, in banking, finance and insurance – geographical locations such as the City of London and Wall Street in New York were vital spatial centres in establishing trust, reputation and expertise through personal contact networks (Storper and Venables 2004). In this context, the geographical origin of such services was situated in a specific place. As specialization in services has proceeded apace, particular service activities have been carved out and packaged into tradeable commodities capable of being delivered over the wire or on-line across space and time (Dossani and Kenney 2007). Face-to-face has now become only one amongst numerous modes of affording the relational proximity for services transactions (Boschma 2005). Wally Olins (2003: 17) cites the example of the branded services of Visa which 'is so impalpable that it's a kind of wraith. It seems to have no provenance: it's quite as much at home in Turkey as Thailand. It takes on the protective colouring of those financial service organisations with which it's associated.' In this context, ascertaining the precise origins of where specific branded services come from and/or are associated with has become much more difficult. Where, for example, is the service of your branded bank account from or associated with? At the headquarters of the main brand owner? In the local high street branch? In the call centre that answers your queries? In the server farm that supports your on-line access?

The ways in which manufacturing and service activities are being combined into packages – what Andrew Sissons (2011: 3) terms 'manu-services' – has similarly complicated considerations of the origin(s) of brands and their branding by bundling together and integrating different kinds of economic activities. Business brands historically renowned for their manufacturing such as Boeing, Nokia and Rolls-Royce now no longer represent themselves simply as sellers of tangible goods such as aircraft, mobile phones and aero-engines. They now communicate their brands as providers of material goods integrated with services such as aftercare, customer support and software into packages articulated as 'solutions', 'outcomes' or 'experiences' (Pike 2009c) (Figure 3.3). Brand guru Wally Olins (2003: 51), for example, quotes a Ford Motor Co. executive in America in the late 1990s claiming that 'the manufacture of cars will be a declining part of Ford's business. They will concentrate in the future on design, branding, marketing, sales and service operations.' Such branded businesses have transformed their activities from just producing tangible things through investment and integration with *in*tangible assets such as brand equity as well as design, human

Figure 3.3 Boeing's 'Manu-services'. Source: Boeing.

capital, organizational capital, R&D and software. The strategy has been to better utilize specialist knowledge and technology for differentiation and defence against imitation and undercutting by lower cost competition. Recent research estimated that manufacturers in the United Kingdom, for example, invested 20% of their value-added in intangibles, worth £35 billion annually, and this was more than three times higher than the £12 billion investment in tangible assets (NESTA 2011). Establishing the origin(s) of such branded packages of manufactures and services integrated across space and time within global value chains is now a more complex task.

Third, the 'national' identifiers central to the era when '*Country* of Origin' was a dominant marker of brand provenance have since been played down, reworked or even eradicated in specific cases. Changes have occurred as actors have tried to obscure historical associations or dissociate brands from specific national territories in the context of globalization and efforts to build appeal and reach beyond particular national domestic markets. Periodically, and sometimes as a result of industrial concentration:

> the name of the founders or the regional origin (Pittsburgh Plate Glass) have given way little by little to more symbolic names … the names they choose are often neologisms, with a concern for geographic neutrality and a certain tendency towards abstraction … the brand identity has moved away from the reality to the concept. The name no longer

> connotes a founder or a region or the qualities of a product or service. It's indeterminateness is a federating force. (Chevalier and Mazzolovo 2004: 29, 30)

The downplaying of the 'national' has resulted too from the modernization strategies of actors trying to escape from sticky national attributes and characteristics that have adhered to particular brands but are considered commercially problematic or damaging in shifting market situations. In the wake of globalization and post-colonialism, for example, the Imperialist history of Britain and its entanglement with the idea of 'Britishness' has left some unwanted undertones of exploitation, domination and militarism (Goodrum 2005). Such associations have proved problematic and have been actively obscured and reworked by actors seeking to construct and salvage from reworked versions of 'Britishness' more meaningful and valuable connotations such as authenticity, craftsmanship and design (Chapter 5).

In other areas, nationally situated brand names have been the focus of rebranding initiatives and even wholesale corporate name changes. Formerly nationalized and then privatized telecommunications provider British Telecom, for example, became BT following its privatization and rebranding in 1991. British Petroleum transformed into BP in 2001 in a short-lived attempt to position the company in the renewables market (Moor 2007). As part of its internationalization and rebranding as a 'world citizen', British Airways (BA) removed the Union Jack national flag from the livery of its aircraft in 1996 and deployed new designs of a multi-coloured and multi-cultural character to reflect their global aspirations and customer reach (Beverland 2009). BA later reinstated the distinctive national designs following adverse responses. Elsewhere, apart from in specific sectors strongly geographically associated with particular and traditional skills and cultural authenticity, 'the real national brand is in terminal decline, while the fantasy national brand Neutrogena (Norwegian), London Fog (British), Baileys (Irish), Häagen-Dazs (Scandi-wegian) is flourishing' (Olins 2003: 147). Emergent nation states have sought to articulate and project their national modernization and development narratives through close associations with branded goods and services, for example Emirates and Singapore Airlines (Buck Song 2011).

Last, as explained in Chapter 2, there is a belated and growing recognition in marketing that in geographical terms origin can be more than just the national level connection to a '*Country* of Origin'. In a far-sighted acknowledgement, Nicolas Papadopolous (1993: 4) noted in the early 1990s that:

> Products are not necessarily made in 'countries'. They are made in 'places', or *geographic origins*, which can be anything from a city to a state or province, a country, a region, a continent – or the world, in the case of 'global' products. (emphasis added; see also Morello 1993)

Similarly, Stephen Thakor and Rajiv Kohli (1996: 27) see brand origin as 'place, region or country to which the brand is perceived to belong by its target consumers'. Stephen Thode and James Maskulka (1998) too interpret place-based strategies as specific extensions of 'Country of Origin' because geographic designations signal quality in brand equity. Even ostensibly 'global' brands are seen as geographical in seeking to evoke 'world origin' (Papadopoulos 1993: 18). Examples here include branding activities such as the 'United Colours of Benetton' and the naming of specific products such as the Ford 'Geo'. Marketing practices have turned towards the use of more loosely specified notions of 'place' to summon up intangible or softer attributes – aura, feelings, mystique (Papadopoulos and Heslop 1993: xxi) – and connect to the growth of emotionally based brand differentiators (de Chernatony 2010). Indeed, Wally Olins (2003: 143) recognized this blurring and fluidity in his claim that 'attitudes towards the nation and the brands which derive from it are unpredictable, emotional, variable and spring largely from legend, myth, rumour and anecdote'.

While beginning to acknowledge the possibilities of a wider array of spatial connections and connotations in brands and branding beyond the national frame of '*Country* of Origin', the geographies evident in research undertaken especially in marketing remain underdeveloped and underspecified. Ian Phau and Gerard Prendergast (1999: 1984) even claim that:

> As a recognition of the global manufacturing activities, consumers today are 'conditioned' by the fact that a product is unlikely to be made in the brand country. Thus, where a product is made, is irrelevant. Instead, the consumers will channel the evaluation cues towards the bundle of favourable attributes (such as design, style, quality) associated with the 'brand' instead of going through the complex clutter of multi-country affiliations. Further, the design or concept of a product comes from the headquarters or is at least approved by the country of brand origin.

In this view, brands send out 'signals' about *Countries* of Origin that overwrite any changing dimensions of *Country* of Origin such that 'brand origin will not change with a change in manufacturing location' and 'The perceived origin of a brand need not be the same as the country shown on the "made-in" label' (Phau and Prendergast 1999: 1984). Origination subjects such claims of the irrelevance of geographical associations to further conceptual, theoretical and empirical scrutiny. This aim is important given the need to acknowledge the tensions and accommodations between territorial and relational thinking shaping geographical associations in brands and branding.

In early work in marketing, 'place origins' were seen as 'related to cities, locales, regions, areas, states, provinces, continents, trade blocs, and so on' (Papadopoulos 1993: 29–30). When such scalar distinctions have been drawn these have been left largely unspecified and tend to elide different spatial levels. Nicolas Papadopoulos (1993: 16) mixes scales in his discussion of how:

> Origin information can be … related to regions rather than countries (Eurocar rental agency) and sometimes is used as a key *descriptor* for a product category (e.g. Scotch Whisky, British ale, California wines, Bohemian crystal) (emphasis in original).

Other examples include Jean-Noël Kapferer's (2002: 163) idea of the 'local' brand, which does not specify what is meant by 'local' but attaches the label to everything other than (similarly undefined) 'global' brands. Similarly, Jean-Noël Kapferer's (2005: 319) 'post-global brand' is termed 'regional' but is situated at the supranational scale. Michel Chevalier and Gérald Mazzalovo (2004: 101) acknowledge that a 'brand's 'culture' is linked to the original values of its creators – often, the culture of the country, the region or the city where the brand developed: Madrid for Loewe, Sicily for Dolce & Gabbana, Majorca for Majorica, Japan for Shiseido'. This account at least recognizes the different geographies of such links. Sociological accounts too have loosely specified their geographies. Despite acknowledging that 'Brands … remain attuned to the place-based connotations of their goods', Liz Moor (2007: 24) is unclear in deploying 'regional' at the supranational level alongside 'non-national' but not specifying whether this means global or sub-national.

Significantly for origination, the multiplicity and complexity in origin(s) and their geographies affords heightened degrees of flexibility and fluidity for actors in (re)working the geographical associations of brands and branding. Depending upon specific connotations in particular geographical and temporal markets, a singular origin or plural origins can be less easily or obviously discernable and the object of promotion or obscurity by the actors involved. The internationalizing division of labour and recognition of geographies including and beyond the nationally framed '*Country* of Origin' means actors now have greater potential to play up or hide origin cues, selectively constructing and representing origins through branding (Papadopoulos 1993; Thakor and Kohli 1996). The rise of on-line services in the digital era has further reinforced the flexibility of origins in virtual space. Situating origin(s) can still include the national scale but this is only one spatial level amongst other territorial constructs *and* needs to be situated in tension and accommodation with understandings of relational networks too.

In specific cases, brand owners and circulators emphasized the particular origin and provenance of certain components or materials especially in spatial and temporal market contexts where this is deemed commercially meaningful and valuable. The drivetrain of engine and transmission for a BMW vehicle, for example, is explicitly marked as 'Made in Germany'. This branding communicates the meaning and value of the national engineering reputation of Germany and situates it within the tradition of the manufacturing renown of *Standort Deutschland*. This origination is utilized even when the final vehicle may be assembled for sale in Spartanberg, North Carolina, for the American market or Rosslyn, South Africa, for southern Africa. The national frame is emphasized rather than any sub-national reference to the land level and Bavarian ingenuity or urban connection to the city of Munich. Both of which particular geographical associations may struggle to communicate the owner's desired brand meaning and value to wider international audiences. In other cases, the precise origins of the components and raw materials may be downplayed or obscured. Coltan from the conflict-torn regions of the Democratic Republic of Congo, for example, is utilized in the screens of branded mobile phones but forms no part of their origination (Montague 2002). The different actors involved can manipulate brands to insinuate or obscure geographical associations to place in myriad ways: 'the organic company "Tom's of Maine" puts location at the center of the brand in a way that Procter and Gamble ("of Cincinnati") does not' (Molotch 2002: 680). Elsewhere, brand owners and circulators have sought to work with 'hybridization' and play up the 'bi-national', 'multi-country' or even 'global' affiliations of branded goods. The 'Intel Inside' branding utilized by the micro-electronics producer Intel, for example, signals the brand and American high-tech expertise and know-how if not precise geographical origin of the place of production of the microprocessor chip component in a laptop or PC branded for sale by another producer (Norris 1993). Services too have been subject to an internationalizing division of labour and where they are delivered from is coming under scrutiny in similar ways, for example where R&D and software development activities have been outsourced internationally to offshore providers in India (Mudambi 2008).

The socio-spatial histories of brands and branding

As Arjun Appadurai's (1986) social lives and histories of commodities and Igor Kopytoff's (1986) commodity biographies suggest, the temporal aspects of origin are vitally important because branded goods and services and their branding can have long histories that shape their subsequent development in often meaningful and valuable ways (see also Koehn 2001; Room 1998). Over time and through space, the inescapable geographical associations in branded objects and branding processes accumulate what Kevin Morgan *et al.* (2006: 3) call 'social and spatial histories' that, to varying degrees and in differing ways, condition and pattern their future evolution. Recognizing the inescapable geographical associations in brands and branding, the spatial historical dimensions are conceived as territorial *and* scalar as well as relational *and* networked.

Depending upon the life course and longevity of the specific commodities, the social and spatial histories of branded goods and services amass archive collections of materials: outdated models and versions, designs, prototypes, packaging, adverts, brochures, software, discontinued lines and so forth. In the context of cognitive-cultural capitalism, for Allen Scott (2010: 123) such archives comprise:

> a stockpile of knowledge, traditions, memories, and images. These assets function as sources of inspiration for artists, designers, craftsworkers, and other creative individuals. As such, they also leave traces on the final products of these workers, imbuing them with an air of authenticity, and hence also contributing to the logic of product differentiation characteristic of so much of the modern cultural economy.

These repositories are where the brands have come from in terms of origin(s) and where they have been. In some cases, discussed in the extended case analyses of Newcastle Brown Ale, Burberry and Apple (Chapters 4, 5 and 6), such resources provide the basis of traditions and narratives. These stories are periodically distilled and continuously reworked – often highly selectively – by the actors involved in efforts to create and fix meaning and value in certain market times and spaces.

Branding actors often try to construct appealing 'commodity biographies' to articulate stories of origin(s) as well as qualities such as authenticity, provenance and quality that are meaningful and valuable in particular spatial and temporal market settings (see, for example, Hughes and Reimer 2004; Jackson *et al.* 2006). Distinctive identities and histories draw upon geographical associations to establish authenticity and attract, stimulate and sustain differentiation in specific brands in the context of media pluralization and cacophony. Speciality food brands, for example, are often:

> marketed in ways which try to exploit the cultural meanings attached to the region of production. ... In linking products to 'cultural markers' or local images such as landscapes, cultural traditions, and historic monuments, their value can be enhanced because

> consumers come to identify certain products with specific places. (Ilbery and Kneafsey 1999: 2208)

Recognition of the historically evolving and inescapably spatially rooted identities, personalities and narratives crafted by actors for goods and services brands through branding processes are recognized in research in other disciplines too. Examples include the work of branding commentators (Aaker 1997), economic anthropologists (Wengrow 2008), economic historians (Da Silva Lopes 2002) and sociologists (Holt 2006b).

The long histories of brands and branding are important because branded goods and services commodities accumulate geographical associations made up of characteristics, identities and values from which extrication or reworking is difficult. Such socio-spatial histories impart a degree of path dependence shaping – to varying degrees and in different ways – their subsequent evolution, trajectories and branding. In particular market contexts, some brands are unable to shake off such associations. McDonalds' reputation for poor quality fast food, for example, and its links to American economic and cultural imperialism have proved resilient. Such markers have endured despite McDonald's recent brand makeover and attempts to improve the dietary quality of their products in the United Kingdom and elsewhere (Ritzer 1998). Other brands have become tainted by spatial connotations in geographically differentiated ways. Danish products were boycotted in Islamic countries following the religious cartoons controversy in 2006. Shell was forced to dispose of its assets in America following the Deepwater Horizon rig accident and oil spill in the Gulf of Mexico in 2012. In this way, the associations of 'geographical lore' can be sticky, slow changing and adhere to particular branded commodities (Jackson 2004). Practices, even events, may rapidly reshape origin perceptions and contaminate the geographical associations of goods and services brands in particular spatial contexts during specific periods.

Yet such historically accumulated attributes do not simply determine the future development of brands and branding. Branded goods and services commodities are not simply prisoners of history and always and everywhere determined by their past. Perceptions and geographical associations can be actively (re)shaped over time to the commercial benefit of the brand through the agency and branding practices of actors in spatial circuits. This is especially the case where brands are new market entrants. Huawei, the emergent electronics and telecommunications company from China, has engaged in intensive brand building activities. The strategy is to establish and differentiate its brand against international competition in the large and growing domestic market in China and to provide a platform for longer-term internationalization. Huawei has established four new R&D centres in Finland, Ireland, Italy and the United Kingdom, rationalized its product line and proactively built its brand profile (Hille 2013). Such activity in commercial markets has not prevented controversy concerning the relationship between Huawei and the Chinese state that has attracted criticism (Hille 2013). Important here has been changing the reputation of China as a place where brands and branding can be originated rather than just a place that produces brands originated elsewhere. This long-term goal is recognized in national development strategy with the changing emphasis in recent advertising campaigns from 'Made in China' to 'Made in China, Made with the World' (Ryssdal 2009).

Origination

Geographical origin(s) have become more complex. Provenance has moved beyond the idea of '*Country* of Origin' in the context of the internationalizing division of labour. The social and spatial histories of brands and branding are integral and formative in shaping their meaning and value. Origination provides a means to conceptualize and explain how, why, where and in what ways actors construct geographical associations in branded commodities and their branding in trying to imbue meaning and value for specific goods and services in particular spatial and temporal market contexts. The focus is upon how actors wrestle with embodying *and* reacting to the rationales of accumulation, competition, differentiation and innovation that disrupt their attempts to secure in space and time geographical associations integral to fixes of meaning and value in their branded commodities in spatial circuits. Origination means the ways in which geographical associations are constructed by actors for branded commodities and their branding that connote, suggest and/or appeal to particular spatial references that embody and mean certain valuable things in specific market situations. Put simply, for goods and services origination is what actors involved in brands and branding do to try and show or imply where something comes from and/or is associated with as a way of making and presenting goods and services that people want to buy at certain times in some places. Context is critical in understanding how actors originate different goods and services in varied ways in particular market times and spaces. Much more than a descriptive metaphor, origination provides a conceptualization and theorization of the social and spatial processes involved.

The idea of a single geographical origin for a branded good or service commodity has been complicated by the internationalizing division of labour and the recognition of geographical associations beyond the singular scalar frame of the national '*Country* of Origin'. Origination, then, does not attempt or expect to locate or situate in a single space or place any literal kind of essential primary source or genuine form from which others are derived. Such an origin *may* be identifiable for specific branded goods and services, enduring over time and space as an authentic and unique reference point. But, as Celia Lury (2011: 50) points out, there is more often no 'single originary moment' that can be established for a particular branded good or service commodity. Complex divisions of labour and global value chains bring into the frame plural origins for branded goods and services. Rather than some kind of essential entities existing in the ether waiting to be discovered and revealed to the world, the branded commodity and its branding are socially constructed, circulated, appropriated and regulated through the agency of actors in spatial circuits. As explained in Chapter 2, in creating and reproducing the geographies of brands and branding, producers, circulators, consumers and regulators play integral roles. Even where claims by particular actors are made to some kind of authentic or intrinsic origin or origins they warrant scrutiny to establish their veracity. The social and spatial histories of brands and branding are articulated by actors with incentives for myth-making and story-telling in attempts to create value and meaning in their brands (see, for example, Holt 2004).

Defining provenance as 'place of origin', it is illuminating to connect to its Latin root – *provenire* – meaning 'to originate' (*Collins Concise Dictionary Plus* 1989: 1036). At particular moments in spatial circuits of value and meaning, origination is a verb: to *originate*. This conception enables thinking about and explaining how actors attempt to *originate* – bring into being – branded commodities and their branding. Origination is also a noun or adjective to name or describe the precise *origination* of a particular branded good and/or service. This way of thinking privileges the agency of actors in actively *doing* the originating. It affords explanatory weight to spatial and temporal market situations and the imperatives of accumulation, competition, differentiation and innovation that shape and disrupt rather than simply determine the work of brand and branding actors.

Building upon the territorial and relational geographies in tension and accommodation discussed in Chapter 2, origination is variegated. Actors try to construct – wholly or in part – geographical associations of different kinds, degrees and characters in and through branded commodities and their branding as part of efforts to create and fix meaning and value. Celia Lury (2004: 54) engages this diversity, heterogeneity and variety in conceiving of the 'origin-ality' of the brand as a new media object and interface. Taking the sportswear brand Nike, Celia Lury (2004: 55) argues that:

> the origin-ality of the Nike interface is less clearly tied to a single national place of origin, or indeed to an origin at all ... multiple origins for the brand are brought into being ... the interface of the Nike brand ...appears as if there is no need to locate this ethos within territorial boundaries in order to secure its ownership or claim its effects ... the interface is not tied ... it is de-territorialising ... since the brand's origins are not visibly tied to specific places of production, the Nike company is able to exercise enormous spatial flexibility in relation to the place of manufacturing of its products. (Lury 2004: 54–5)

Celia Lury (2004: 55) is not claiming 'that the Nike brand functions without limits' but is seeking 'to show that the performativity of the interface is such that *the relation of a brand to an origin may be organised in many different ways*' (emphasis added). Such relational thinking can deepen explanation of origination when married with a territorial sensibility. Other critical work on Nike, for example, has emphasized the brand's connection and transmission of a territorially rooted national American culture and the territorial boundaries of jurisdictional authorities regulating the intellectual property rights underpinning Nike's trademark protection against counterfeits in international markets (Goldman and Papson 1998).

In the conceptual frame of origination, we can interpret actors trying *to originate* specific branded goods and services commodities in particular ways. Different kinds, varying extents and differing natures of geographical associations are used to summon up and hold onto meaning and value in certain market situations (See Table 2.6). At one end of this view, strong, explicit and direct forms of origination are attempted by actors to construct and appropriate meaning and value. With intrinsic and regulated geographical associations to particular places, agro-food brand producers often make claims of the authenticity of enduring and longstanding provenance. The concept of *terroir* is an example in wine geographies (Banks *et al.* 2007). At the other end, actors may work the ambiguity, fluidity and even ignorance and lack of knowledge of

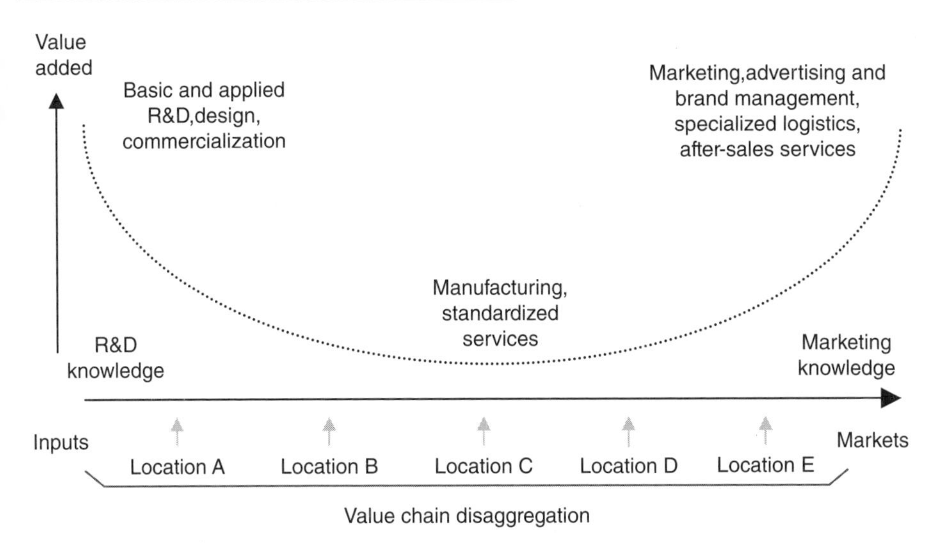

Figure 3.4 'The smile of value creation'. Source: Adapted from Mudambi (2008: 707).

provenance. On-line interaction in virtual space accelerates the pliability of origin claims. Actors construct weak, implicit and indirect forms of origination to obscure or entirely invent geographical associations deemed capable of meaning making and value creation for certain branded goods and services commodities. Demonstrated in the extended case analyses (Chapters 4, 5 and 6), this kind of origination can occur for existing brands where particular spatial circuits have undergone substantive change such as disconnection and/or relocation and specific forms of geographical association have been modified and/or transformed. Actors launching new branded goods and services commodities can attempt to write their own spatial scripts from scratch and make appeals to assorted geographical associations and imaginaries in bids to create value and meaning in their socio-spatial markets. Kingfisher lager's 'Authentically Indian' branding in the United Kingdom, for example, directly situates the brand through associations to the Indian sub-continent. This strategy is part of constructing its meaning and value as an accompaniment to Indian cuisine in restaurants and takeaways while simultaneously downplaying the particular nature of its ownership and brewing activities under contract to the Netherlands-based group Heineken in the United Kingdom.

Building a stronger appreciation of geographies into 'country of origin of brand' (Phau and Prendergast 2000: 165), as specialization and the division of labour extends geographically for goods and services commodities, origination is being distinguished by especially brand producers and circulators for specific parts of their spatial circuits of value and meaning. The flight from cost sensitive assembly and production in more advanced economies has underpinned the outsourcing of assembly, production and increasingly service delivery activities to lower cost suppliers of comparable productivity and quality internationally (Dossani and Kenney 2007). Figure 3.4 illustrates what Ram Mudambi (2008: 706, 2007) argues shapes the geographical patterns of control and location in value chains internationally because:

> firms are finding that value-added is becoming increasingly concentrated at the upstream and downstream ends of the value chain. ... Activities at both ends of the value chain are intensive in their application of knowledge and creativity. Activities at the left or 'input' end are supported by R&D knowledge (basic and applied research and design), while activities at the right or 'output' end are supported by marketing knowledge (marketing, advertising and *brand management*, sales and after-sales service). ... The geographic realities associated with the smile of value creation are that the activities at the ends of the overall value constellation are largely located in advanced market economies, while those in the middle of the value chain are moving (or have moved) to emerging market economies (emphasis added).

Economic imperatives have compelled actors in especially higher cost and higher wage economies to develop more sophisticated, productive and higher value-added activities to establish defensible and distinctive market positions against lower cost international imitation and replication (Storper 1995). Simultaneously, actors from emerging economies such as Brazil, China, India and Mexico are seeking to upgrade in the hierarchy within global value chains. Activity is focused upon the development of service and R&D inputs to basic production and assembly as an Original Equipment Manufacturer (OEM) in distribution and purchasing through Original Design Manufacturer (ODM) to the pinnacle of 'Own Brand Manufacturer' status (Humphrey 2004). The upgrading rationale moves actors 'to develop their own brands and marketing expertise in advanced economies to increase their control over the downstream end of the value chain' (Mudambi 2008: 708).

In 'cognitive-cultural capitalism' (Scott 2007: 1465), the narrative and policy support for transition towards the 'knowledge economy' and emphasis upon innovation, design and styling have reinforced and enabled processes of specialization and geographical expansion of the division of labour. Indeed, discussed further in Chapter 7, such currents have further established the potency and reach of the idea of the brand and the process of branding beyond the world of goods and services commodities into the sphere of territorial development. Dominic Power and Atle Hauge (2008: 3) even argue that branding has become 'one of the core strategic and commercial competences driving firms, clusters, regions, and nations in the contemporary economy'. Such developments in branded goods and services markets have not proceeded in a simple and geographically uniform fashion towards what Naomi Klein (2000: xvii, 4) described as the 'post-national vision' of a more 'de-materialized', 'weightless' economy in which what 'companies produced primarily were not things ... but *images* of their brands' (emphasis in original). Considering and reflecting upon the origination of brands and branding presents a more geographically nuanced picture. Origination affords a means of conceptualizing and capturing the more complex variety and diversity evident amongst the actors constructing and responding to the contemporary geographies of brands and branding.

The conceptualization of origination of goods and services brands and their branding demonstrates how marketing's influential notion of 'Country of Origin' – and the more recent 'country of origin of brand' – has now fragmented and splintered. The globalizing division of labour has meant the spatial dispersion and integration of the 'trade in tasks' rather than only trade in completed goods or services (Kemeny and Rigby 2012). Analytically, ten – perhaps more – different categories of origination with which actors in spatial circuits are working can be identified:

1. where the brand was originally born;
2. where the brand was innovated or thought up;
3. where the brand is designed and developed;
4. where the brand is tested and refined;
5. where the brand is headquartered (or indirectly where the brand owner is headquartered);
6. where the branded good or service is delivered or provided from (physical, virtual);
7. where the brand is made (manufactured and/or assembled);
8. where the brand is sold (wholesale or retail channels);
9. where the branded good or service is serviced or supported from (e.g. after-sales services, maintenance);
10. where the brand is recycled and/or disassembled.

Further, the order of importance of such forms of origination has shifted too. As Chris Moore, Professor in Marketing and Retailing at Glasgow Caledonian University, noted:

> place of manufacture, while initially was the prominent one, has become the least important one. ... People start to say what matters more, is it the design idea and the ingenuity, or is it the place of making. And it seems for a lot of people it is the idea. (author's interview, 2008)

In the context of complex and internationalized global value chains, John Quelch and Jennifer Jocz (2012: 44) too argue that 'there remain product categories where it pays marketers to ensure that superior expertise is still widely associated with a particular place' and 'the power of brand trust is expected to override any doubts customers might have as a result of products being sourced from multiple countries. Consumers are more concerned about the country or place of design and quality oversight than the country of place of manufacture.' As a way of explaining what actors do in attempting to summon up meaning and value in goods and services commodities through geographical associations, origination is an emergent and fruitful idea. What origination needs is empirical exploration and challenge, examining its robustness and worth in understanding the geographies of brands and branding. Prior to its more detailed examination in the extended case analyses of the brands and branding of Newcastle Brown Ale, Burberry and Apple (Chapters 4, 5 and 6), it is important briefly to elucidate the complex and emerging ways in which origination is evident. The clothing industry and tele-mediated services provide examples for further and brief elaboration and interpretation to begin to examine the worth of origination as a means of conceptualization, theorization and analysis.

Origination in clothing and tele-mediated services

The clothing industry has a long history of internationalization and a geographically expansive division of labour marked by the enduring importance of brands and branding as well as provenance as a sign of meaning and value (Dicken 2011; Dwyer and Jackson 2003). Contemporary examples demonstrate how actors are originating their brands in particular ways. US-based American Apparel, for example, foregrounds 'Made in Downtown LA' as integral to its business ethos and brand and uses it to mark its products and retail outlets. The actors involved differentiate the brand in competition as a Sweatshop-Free' and vertically integrated in contrast to the low cost, vertically disintegrated and international sub-contracted business models prevalent in the clothing industry (Ross 2004). The brand is presented as an 'Industrial Revolution' with its key R&D, marketing and manufacturing activities located in downtown Los Angeles. This narrative is articulated in the brand's circulation and consumption in its retail outlets and in its website which invites potential consumers to 'Explore our Factory'. Origination enables interpretation of the geographical associations constitutive of the meaning and value of the origination of the brand. The actors involved have originated the good and service commodity in a nationally situated brand name ('*American* Apparel') and articulated a 'Made in…' claim that is located not simply at a national level but in a specific territory within a particular city – the downtown area of the City of Los Angeles, California. As a place, LA resonates as a centre of innovation, style and buzz in global fashion circles (Tokatli 2013). Origination explains the particular kinds and forms of geographical associations central to the brand's identity and its branding activities.

In common with American Apparel, origination reveals how the owners and circulators of the Skunkfunk clothing brand mix a geographical association for a specific function in a particular sub-national territory – 'Designed in the Basque Country' – with acknowledgement of the provenance of its manufacture in a national 'Made in…' claim – 'Made in China'. With this origination, the actors are constructing meaning and value by locating its roots and source ideas in the culturally rich, meaningful and valuable traditions and style of an authentic (sub)national place (the Basque Country). In addition, the actors have had to acknowledge its international outsourcing (to China) in the context of regulatory controls on product labelling. In contrast to the territorial forms of origination for the brand examples of American Apparel and Skunkfunk, the actors involved in the clothing brand Superdry deploy a more relational form of origination. The brand makes a 'Made in…' claim in relation to a specific territory (China) alongside an emphasis upon 'British Design. Japanese Spirit'. This kind of origination melds particular activities in specified territories at the national level – 'British Design' – with the suggestion of distanciated and relational influence and inspiration from another specific national territory – 'Japanese Spirit'. Origination, then, can conceptualize and explain the tensions and accommodations between territorial and/or relational origination of brands and their branding.

Origination demonstrates its conceptual and theoretical worth in helping to explain more complex forms of geographical association too. In the context of 'Made in…' claims becoming more complicated and being challenged, the enduring and much vaunted meaning and value of 'Made in Italy' in the clothing industry has been questioned by the activities of brand and branding actors. Italian fashion houses have been accused of outsourcing some of the more basic stages in the manufacture of branded goods to lower cost contractors in Bulgaria and Romania in central and eastern Europe as well as China, then undertaking only the finishing stages in Italy to substantiate the 'Made in Italy' label and premium price (Thomas 2007, 2008; see also Hadjimichalis 2006; Ross 2004). Amidst debate about legal regulations aiming more tightly to control the 'Made in Italy' label, the fashion house Prada launched its own 'Made in...' range. The actors emphasized the authentic provenance of particular designs made by Prada while drawing inspiration and materials from *and* being produced in particular national territories. The branded goods bore the marks 'Prada, Milano' and the labels 'Made in India', 'Made in Japan', 'Made in Peru' and 'Made in Scotland' (Figure 3.5). In the midst of the controversy, Miuccia Prada declared:

> 'Made in Italy? – who cares? It's not brand strength if you have to defend your work. Mine is a political statement and it comes from a personal appreciation of originality. You have to embrace the world if you want to live now. ... It's taking away hypocrisy. ... In Europe we have to retain values – there is so much history. But how can it survive, unless there is a contemporary way to show our history in a way that young people get it? (Miuccia Prada, 2010 quoted in Menkes 2010: 1)

In this example, Prada is deploying a hybrid form of origination. The actors are combining a longstanding geographical link between the brand and the specific territory of the city of Milan in Italy with a nationally situated origination differentiated by the specific type of branded commodity and its constituent designs and materials. Origination enables a critical engagement with brand actors try to construct, fix and appropriate geographical associations in pliable and sophisticated ways. The approach demonstrates how actors create meaning and value using particular geographical imaginations. It provides a way to understand the uneven geographies of investments, jobs and territorial development underpinning brands and branding.

Origination does not just demonstrate its interpretative and explanatory worth in relation to the manufacture of branded goods: branding actors deploy geographical associations in efforts to create and fix meaning and value in branded services too. In tele-mediated services, the origination of where the actual tasks are undertaken and services delivered from has become more complex. New socio-spatial divisions of labour have begun to influence how branding actors originate their services, inflecting the spatial connections and references articulated in trying to construct meaning and value in market situations. The standardization and routinization of back-office service functions has led to the development of contact or call centres and their decentralization to economically lagging places with pools of abundant lower wage, often women, workers (Richardson *et al.* 2000). Competition and cost pressures encouraged outsourcing and the international offshoring of such contact centres facilitated by advances in information and communication technologies (Dossani and Kenney 2007). However, problems with quality, customer disquiet and competition

Figure 3.5 Prada 'Made in…' labels. Source: Prada SA.

Figure 3.6 'Just returning your call … to the UK'. Source: http://news.bbc.co.uk/1/hi/magazine/6353491.stm, accessed: 19 November 2014. Reprinted with permission of the BBC.

have led some service providers to return their contact centre services to the United Kingdom via both in-house and external contractors. This process is part of a wave of what has been termed 'on-shoring' (Christopherson 2013) (Figure 3.6). Such changes in economic organization have influenced the articulation and circulation of the origination of branded services. The bank NatWest, for example, used a recent advertising campaign to bolster its brand's reputation for 'Helpful Banking' and question its competitors by asking 'We have award-winning 24/7 *UK* call centres. Does your bank?' (emphasis added). The branding actors have explicitly constructed their services as higher quality but equally accessible because of their origination in '*UK* call centres'. The emphasis upon the United Kingdom here is central to the meaning and value of the brand's proposition *vis-à-vis* its competitors who may still be offering services outsourced to external – and by implication lower quality – providers internationally. NatWest is explicitly promoting the origination of its service delivery and its connection with consumers anxious to access services from the specific territory of the United Kingdom. This origination is being deployed even when the actual services are delivered through tele-mediated means over the wire and could technically be provided from virtually anywhere and certainly beyond the United Kingdom.

The origination of branded services commodities is not just evident at the lower value end of activities. In higher value-added services such as software engineering, issues of where services are delivered from have emerged as part of the growth in offshoring in recent years. Existing producers facing cost and competition pressures in advanced countries have sought international outsourcing for activities that are modularized, standardized, amenable to automation, have low barriers to entry and do not require regular contact with consumers (Dossani and Kenny 2007). Established brand name firms have been offshoring software activities for both in-house product development, including TI, Agilent, HP, Oracle and GE and service delivery systems, including ANZ Bank, ABN Amro Bank, Accenture, IBM and Dell. New producers from India and externally owned branch operations have been established in emerging economies to cater for this demand, supported by national industrial and regional development strategies and the 'brain circulation' of skilled labour across the Pacific Rim (Saxenian 2005). Concentrated geographically in centres such as 'The Silicon Valley of India' in Bangalore, growth has been exponential amongst software firms producing for export markets. Employment has expanded from 27 500 to 513 000 between 1995 and 2005, while the number of firms grew from 816 to 3170 between 2000 and 2004 (Dossani and Kenney 2006). Silicon Valley firms, such as Tensilica, have actively shifted 'second generation' work including adding features to existing software and improving reliability to India while redeploying more expensive US-based software engineering resources into new product development (Dossani and Kenney 2006: 14). Such developments have served to counter the initial scepticism about contracting-out higher value-added activities to producers in emerging economies and dispel the assumed links between low cost, low productivity and low quality. However, the brand name firms in the advanced economies have had carefully to manage how this process and its outputs are originated and how it is perceived by their international customers because 'services exports from developing countries initially lack brand value and are quite different from services exports from developed countries' (Dossani and Kenney 2006: 18). In Tensilica's case, the work was initially contracted with a vendor to save on set-up costs with the aim of transferring to a fully owned subsidiary in due course. The early operations were hampered by concerns 'that it lacked a brand name in India to be able to recruit the best talent' (Dossani and Kenney 2006: 14). The software industry in India has had to wrestle with the enduring problem that national goods and services from 'developing' and poorer countries tend to be assessed less favourably than those imported from 'developed' and richer countries (Phau and Prendergast 1999). This is especially the case when those imported goods and services carry a reputable and well-known international brand name. However, as the Indian software industry has grown and developed, its own indigenous producers are beginning to establish brand name and reputation within international markets, including TCS, Infosys and Wipro.

These examples of clothing and service commodities begin to demonstrate the worth of origination in understanding and explaining how actors involved in brands and their branding are balancing strong geographical associations with valuable and meaningful places and spatial connotations of anywhere more commercially ambiguous or even damaging. The originations attempted by such actors are not only and simply fixed at pre-determined territorial scales of origination – 'local', 'regional', 'national' or 'global'. They are constructed by actors in more fluid, shifting and ambiguous ways, moving within, between and across scales within relational circuits and networks. It is vital in challenging the robustness of the theorization, to examine empirically how origination is being grappled with by brand and branding actors by uncovering their particular socio-spatial histories and what producers, circulators, consumers and regulators have been up to.

Researching origination in the geographies of brands and branding

The conceptual and theoretical enquiry of origination raises questions of research method, design and analysis. How can we study origination? How can we build upon and move beyond the single case studies of the geographies of specific brands and their branding that have dominated much research to date (Pike 2011d)? What kinds of plural and mixed methods and comparative frameworks are appropriate? How can we distil more generally applicable analytical frameworks from particular experiences that enable refinement of the conceptualization and theorization of origination? Given the complexity, diversity and variety in the world of brands and branding, Martin Kornberger (2010: 271) rightly makes the point that 'only detailed empirical research can explore how brands are enacted, played out and appropriated in different local contexts'.

Recognizing the integral importance of history and situating it in geographical context, Douglas Holt's (2006b: 359; 2004) genealogies method comprises 'detailed analysis of the social construction' of brands and branding. This approach enables the construction of socio-spatial biographies of particular branded commodities to uncover their 'social and spatial histories' (Morgan *et al.* 2006: 3). This method draws from understandings of the social lives and histories of commodities (Appadurai 1986), commodity 'careers' (Kopytoff 1986: 66) and material culture ethnographies (Miller 1998), as well as geographical 'biographies' and 'lives' (Watts 2005: 534), 'commodity stories' (Hughes and Reimer 2004: 1) and 'life stories' (Smith and Bridge 2003: 259). A historical approach to studying brands and their branding is important. The brand story and narrative of origin are integral to the efforts of actors to create and fix meaning and value for:

> brands that have become successful on the global stage ...have stories spanning several decades. Whether publicized or not, the origins of these brands are original and compelling ... people are attracted to brands that are true to their origins. A strong heritage is not only a sign of authenticity but also a sign of success. People recognize and respect an authentic brand that was created by a specific person or persons and has stood the test of time. (Hollis 2010: 62)

History is important too because of the integral role of guarantee and reputation in sustaining the value and meaning of brands over time and space: "'The prolongation of a contract in time is intrinsic to the notion of guarantee. In order to exist, a brand must not only establish its reputation, but establish it durably. Thus the chronological perspective is fundamental to understanding brands' (Chevalier and Mazzolovo 2004: 15).

Socio-spatial biographies are constructed through extended case analyses of specific brands and their branding. The extended case analysis illuminates the value of the wider claims of origination as a conceptual and theoretical framework rather than just documenting the particular and contingent detail of specific brands and their branding. The example brands and their branding are selected as a 'critical case' to challenge the conceptualization and theorization of geographical associations and origination with the aim of 'generating new theoretical insights, rather than merely illustrating extant theory claims' (Barnes *et al.* 2007: 10). Socio-spatial biography uses mixed methods to trace the different kinds, extents and natures of geographical associations being deployed in efforts to originate meaning and value in brands and branding in spatial circuits (see, for example, Pike 2011a). The accumulated, layered and changing nature of socio-spatial histories create often dense constructions that make it difficult and time consuming to analyse the precise form, degree and character of geographical associations central to the meaning and value of the brand and its branding. This is especially important and challenging given the historical character of processes of brand construction and branding, requiring access to historical sources to track longitudinal change over time and space. As Allen Scott (2010: 123) put it, 'The association between place and product …tends to be self-reinforcing over time because both of them are joined together in a spiral of mutual interdependencies built upon the creative reprocessing of old images and the continual addition of new ones to local repertoires of designs and symbologies'.

As a conceptual framework for analysis and comparison, the socio-spatial biography is situated within a spatially sensitive 'full circuit of capital' (Willmott 2010: 517). This view distinguishes specific but closely interconnected 'moments' of not only production and consumption but circulation and regulation (Hudson 2005; Smith *et al.* 2002). As explained in Chapter 2, this analysis considers the different kinds of socially related and geographically situated actors, activities, relations and processes within the context of the broader and interconnected socio-spatial circuit (see Table 2.10). While it is acknowledged that a blurring between the roles of specific actors in creating meaning and value is taking place in some settings (see Chapter 2), the distinctions between different types of actors are retained to support analysis and interpretation. Second, this approach emphasizes and enables dynamic analysis of how meaning and value have constantly to be produced, circulated, consumed and regulated in branded commodities by the actors involved in coherent and stable ways on an ongoing basis for the expansion of capital to be 'routinely reproduced' (Hudson 2008: 424). The spatially differentiated and uneven geographies of brands and branding in origination are explained, then, by a 'focus upon flows of value and the differential power and position of economic actors in the governance of these flows' which 'allows for an understanding of *which actors* and *which places* benefit from or lose out from such flows' (Smith *et al.* 2002: 54; original authors' emphasis).

Reflecting the tensions and accommodations between territorial and relational understandings of geographical associations, the complex, even contested, nature of the 'placeness of the brand' (Molotch 2002: 679) means identifying a 'whole host of specific interconnections … to be traced across a variety of historical/geographical scales' (Cook and Harrison 2003: 311) as well as spatially extensive flows, surfaces and networks (Lury 2004). The approach here responds to Peter Jackson's (2004: 173) call for more empirical work on brands beyond the relatively narrow range of goods and services studied to date, such as clothing and food (e.g. Cook and Harrison 2003), and different worlds of brands and branding from the well-studied global examples, such as MacDonalds, Nike and Starbucks (e.g. Goldman and Papson 1998; Klein 2000; Ritzer 1998). The framework developed is a first step in attempting to interrogate systematically and comparatively the analytical and explanatory purchase of origination and trace the diversity and variety of complex and overlapping geographical associations constructed by actors in brands and branding.

Specifically, the socio-spatial biography method and research design pieces together the socio-spatial histories of the selected brands and their branding. The research, analytical strategy and account are organized around the moments of the spatial circuit: production, circulation, consumption and their regulation. The actors were identified from the conceptual categories (producers, circulators, consumers, regulators) and selected through review of secondary sources and snowballing to accumulate contacts within wider networks. Secondary sources were especially important given the limited access granted by the high-profile Burberry and Apple brands in these guarded businesses in which brand image and reputation are critical to commercial prospects and tightly controlled (Tungate 2005). As Jan Lindemann (2010: 152) notes 'brand specific information reported by their corporate owners is very scarce'. Apple, in particular, 'has a reputation for secrecy' (Duhigg and Bradsher 2012: 3).

Over the period 2003 to 2013, the empirical research activities comprised, first, the collation and analysis of secondary sources. The empirical material collated, analysed and triangulated comprised published and unpublished materials. The sources included: trade, financial and general media (e.g. *Campaign*, *The Financial Times*, the BBC); articles and reports (e.g. academic, policy and think-tanks); presentations (e.g. annual reports and accounts, financial results, share sale prospectuses); correspondence (e.g. with corporate and other institutional officials); and, images (e.g. adverts, photos). Second, where access was granted, 49 semi-structured interviews were undertaken with key research subjects involved in the brands and their branding. Interview subjects included a wide array of public, private and civic actors ranging across the production, circulation, consumption and regulation moments of their spatial circuits. Where utilized as quotes in the text, empirical material from the interviews has been anonymized and referenced as 'respondent type, author's interview, year'. The interviews sought 'close dialogue' (Clark 1998: 73) with actors to gather evidence systematically and examine the explanatory value of geographical associations and origination in the construction and stabilization of meaning and value in the brands and their branding. The key concerns addressed included: how and by whom the brand and its branding were created and constructed; the role of geographical associations in efforts to articulate, cohere and stabilize meaning and value; the identification and roles of the key actors involved in the spatial circuits for the branded goods and services; how the particular versions of origination in each brand were being held together in the face of accumulation, competition, differentiation and innovation; and the tensions in each brand's spatial circuits between territorial impulses tying the brands to specific areas and relational networks dragging and stretching the brands across space and time.

For the analysis of origination, Newcastle Brown Ale, Burberry and Apple were selected to construct their socio-spatial biographies and analyse how the actors involved in their brands and branding sought to create meaning and value through geographical associations and origination. The three brands were chosen to provide a comparative view on origination by exploring different kinds of brands and their branding in different businesses amongst actors working with common challenges of constructing and stabilizing meaning and value through geographical associations. Each set of actors was involved in ongoing attempts to originate their branded goods and services commodities in the context of accumulation, competition, differentiation and innovation pressures in different and changing spatial and temporal market situations. The research strategy and empirical analysis builds upon and extends the rich insights provided by the single case study, for example Dominic Power and Atle Hauge's (2008) and Nebahat Tokatli's (2012a) studies of the Burberry brand and its branding.

Each extended case study of the brand enabled the examination of the territorial and relational tensions and accommodations shaping the particular origination of the geographical associations of the brands and their branding. The analysis was situated within *and* extended beyond the particular spatial references framing their origin(s): the '*local*' of the city of Newcastle upon Tyne and north east England for Newcastle Brown Ale in the international brewing business; the '*national*' of Britain and 'Britishness' for Burberry in the international fashion business; and the '*global*' of worldwide reach and resonance for Apple in the international technology business. The socio-spatial biographies are not entirely comprehensive and exhaustive; this kind of account would reach beyond the format of a single book. A fuller origination analysis of each brand would warrant book length treatments in its own right (see, for example, Griffiths 2004).

Newcastle Brown Ale (NBA) was selected as an iconic brand – what Douglas Holt (2006b: 357) defines as a 'prominent and enduring cultural symbol' – with historically strong and deeply embedded but changing geographical associations to the particular 'local' of the city of Newcastle upon Tyne and the region of north east England in the United Kingdom. NBA is synonymous internationally with the city and takes part of its brand name from its origin. NBA's socio-spatial biography sought to explain the shifting origination of the brand during its market survival in the United Kingdom and growth in America (Chapter 4). Burberry was selected as a 'quintessential British brand' (Goodrum 2005: 20) in the international fashion business. Connecting political *with* cultural economy in spatial circuits, brands and branding are 'nowhere more important than in the fashion industry, where style and image are as important in the creation of exchange values as textiles and labour power' (Dwyer and Jackson 2003: 270). Burberry's brand owner has sought to (re)construct, cohere and stabilize geographical associations in the 'national' frame of the 'national imaginaries' (Reimer and Leslie 2008: 145) of 'Britishness' for its differentiated meaning and value in changing spatial and temporal market contexts. Apple was selected as a 'global brand' (Hollis 2010: 25–6) with worldwide size, profile and reach. The brand has 'transcended its cultural origins to develop strong relationships with consumers across different countries and cultures' and appeals 'to a relatively homogenous audience' and has 'more of a need to maintain a common image' (Hollis 2010: 25–6). The socio-spatial biography of the Apple brand analyses how the participant actors have attempted to utilize globally the meaningful and valuable attributes and imaginaries associated with its specific geographical associations in the technological heartland of Silicon Valley in California.

Summary and conclusions

Origination conceptualizes and theorizes how, why, by whom, where and in what ways geographical associations are deployed selectively by brand and branding actors to create and fix meaning and value in branded goods and services commodities in the times and spaces of market settings. First, geographical origin(s) and provenance were explained as historically enduring characteristics – part of the inescapable geographies of brands and branding established in Chapter 2 – linking commodities to place and enabling actors to attribute tangible and intangible qualities such as authenticity, reliability and quality in particular market contexts in time and space. Second, the national frame of '*Country* of Origin' has been complicated and questioned by: the internationalizing division of labour and more complex, sophisticated forms of coordination and integration within global value chains; the decline of singular and emergence of multiple and/or hybrid origins; the rise of services and integration of tangibles and intangibles; the downplaying and/or transformation in the use of national identifiers; and the growing recognition of geographies beyond the national in the origin(s) of goods and services commodities. Third, socio-spatial histories explained how the temporal aspects of geographical origin provide resources with which branding actors work, albeit highly selectively, in articulating, constructing and projecting the attributes, characteristics and facets of brands in market situations. Fourth, origination was defined as the ways actors in spatial circuits construct geographical associations in branded commodities and their branding in trying to imbue meaning and value for specific goods and services in particular spatial and temporal market contexts. Variegation is evident in how actors seek to originate identity and worth in branded goods and services commodities, deploying different kinds, extents and natures of geographical associations. Origination advances our understanding by demonstrating how 'country of origin of brand' has splintered into an array of categories, from where brands were originally born to where branded goods or services are being provided. Last, researching origination outlined the method of socio-spatial biography to trace the social and spatial histories accumulated by brands and their branding and explain their origination. Research strategy, activity and analysis focused upon the piecing together of socio-spatial biographies of the case brands using mixed methods to collate, corroborate and triangulate secondary and primary data sources. Newcastle Brown Ale, Burberry and Apple were selected because the branding actors involved have grappled with originating meaningful and valuable geographical associations in particular territorial and relational constructions of the 'local', the 'national' and the 'global'.

Chapter Four
'Local' Origination . . . Newcastle Brown Ale

Introduction

This chapter examines the actors involved in attempting to create and fix meaningful and valuable geographical associations in the brand and branding of Newcastle Brown Ale (NBA) in particular spatial and temporal market settings. Emphasizing the importance of socio-spatial history, it begins by tracing the geo-historical origins of the brand. It explains the origination of NBA in the particular 'local' of the city of Newcastle upon Tyne and the north east region of England. NBA's brand owner, managers, circulators, consumers and regulators have sought to construct and appropriate meaning and value from its traditions and values in its urban and regional commercial heartland and subsequent national distribution. This particular origination made it difficult for the brand owners and managers to recruit new generations of consumers in the north east and beyond in the context of broader market shifts and segmentation. Origination illuminates how the brand owner's search for new markets for the brand articulated with a growing market segment in America. NBA's origination was then changed to the national scale within the pluri-national state of the United Kingdom. The brand was (re)originated as 'Imported from England' in efforts to construct its premium meaning and value for its new college-educated, typically male and affluent younger consumers. The wider value of origination is demonstrated in analysing the particular 'local' origination of NBA to how strong material, symbolic and discursive kinds of geographical associations in brands to particular places can lose their commercial value and meaning over time. Origination illuminates how

Origination: The Geographies of Brands and Branding, First Edition. Andy Pike.
© 2015 John Wiley & Sons, Ltd. Published 2015 by John Wiley & Sons, Ltd.

actors try to reconstruct brands in new spatial and temporal market contexts through subtly different geographical associations.

Producing the 'local' in Newcastle Brown Ale

Geographical associations of the particular 'local' of the city of Newcastle upon Tyne and north east England are evident in NBA's specific origins in Colonel James H. Porter's experiments to develop a distinctive, full flavoured bottled ale brand for Newcastle Breweries in 1927. Facing competition from Nottinghamshire's Burton upon Trent ales 'coming up the country in bottles' (former Marketing Director, author's interview, 2008) and connecting production to marketing from the outset in its spatial circuit, the 'production of difference' (Dwyer and Jackson 2003: 271) sought to create a brand distinct from the commodified, high volume and low profit margin ales and beers available in north east England in the late 1920s (Bennison 2001). The new dark ale was designed to offer consistent quality, taste, higher alcohol by volume, an attractive aesthetic and presentation, and be capable of commanding a premium price. The bottled packaging aimed to 'allow the beer to travel' (former Marketing Director, author's interview, 2008) in a consistent form and quality to support wider geographical distribution and sales beyond Tyneside. With relatively rudimentary technology, initially inherent properties of the new brown ale and its brewing process established NBA's intrinsic and material geographical associations to the Tyne Brewery site in Newcastle upon Tyne. These attachments imbued an origin myth of distinctive 'waters of the Tyne' (Brewery Manager, author's interview, 2008) providing distinctive water combined with locally particular yeast strains and raw materials of barley, hops and malt brewed with locally idiosyncratic and variable brewing equipment and brewers' skills. In this spatial and temporal market setting, Newcastle Breweries had managed to establish and cohere a fix of strong geographical associations to construct a meaningful and valuable 'local' origination for NBA.

Until 2008, the NBA brand was owned by UK-based Scottish and Newcastle (S&N) plc. S&N was growing through acquisition from a regional brewer and pub owner in Scotland and northern England into an aspirant international brewer in the context of a consolidating industry, dominated by Anheuser-Busch/InBev and SABMiller, prior to its £7.8 billion takeover and break-up by Carlsberg and Heineken (Table 4.1). Changing into a brand-oriented sales and marketing company to escape the high capital intensity and sunk costs of brewing, S&N's core strategy in the UK market prioritized high volume 'drive brands' (e.g. John Smiths, Fosters and Kronenbourg) as aspirant market leaders, and higher price and margin 'premiumization' in lucrative 'specialities' niche brands: 'Our brand strength is the key to both growing value in mature markets and rapidly expanding our positions in developing markets' (S&N 2006: 6). Managing S&N's 'federation of brands' (Corporate Affairs Director, author's interview, 2007), non-core functions and 'heritage' brands still valuable in specific market territories were spun-off in joint ventures (e.g. Theakstons) or contracted out (e.g. brewing, packaging). Due to its distinctiveness, uniqueness and international profile for S&N, NBA remained as a 'survivor' (Corporate Affairs Director, author's interview, 2008).

Table 4.1 International brewing groups ranked by output, 2012–2013[a]

Company	*Volume (M/HL)*	*Revenue (US$)*	*Headquarters*	*Ownership structure*
Anheuser-Busch InBev	403	39 758 000	Leuven, Belgium	Public
SABMiller	306	34 487 000	London, UK	Public
Heineken	171	23 686 000[a]	Amsterdam, Netherlands	Public/family Control
Carlsberg	120	11 606 000[a]	Copenhagen, Denmark	Public/foundation Control
China Resources Enterprise	106	8 252 000*	Hong Kong, China	Public

Note: [a]Converted from €, DKK & HKD in 2013. Nominal prices.
Source: Calculated from AB INBEV (2012), Heineken (2012), Carlsberg (2012) and SABMiller (2013).

Oligopolistic rivalry in the brewing business propelled advances in brewing technology, configuring standardized industrial practices in operating new generation 'Brewfactories' with tighter performance management systems. Each of S&N's brands underwent microbiological refinement and codification to a tightly specified recipe and process to maintain their integral characteristics by delivering predetermined attributes of colour, flavour, smell, strength and taste. For NBA, this included: water matching from sources including Northumbria Water's Whittle Dean reservoir; utilization of F40 generic yeast; particular varieties of barley, hops, malt and flavourings (e.g. caramel, hopscotch); and, pasteurization to extend shelf-life. Such practices loosened and, in time, disconnected any intrinsic geographical associations of NBA to the Tyne Brewery site, the territory of the city of Newcastle and the region of north east England until it reached the point at which there was 'nothing really apart from the heritage to keep it on Tyneside' (Corporate Affairs Director, author's interview, 2007). This break freed the brand from any residual spatial ties and rendered its production mobile, allowing S&N to replicate NBA's material attributes beyond the particular geographical and historical 'local' of its initial origination.

Tied into the capital markets in the City of London by institutional ownership as a publicly limited company (PLC) and in the context of financialization (Pike and Pollard 2010), the economic imperatives of scale, specialization and reducing its cost base in a concentrating industry eventually forced S&N to sever the material geographical associations of NBA production in the Tyne Brewery in Newcastle upon Tyne. Pressures from shareholders and analysts in the City for brewing capacity rationalization to reduce its costs and increase investment returns led S&N to close the Tyne Brewery in 2004. The explanation focused on its underused capacity, high costs, ageing plant, limited flexibility, site redevelopment opportunities and need for investment in existing 'drive' brands (S&N 2004). Estimates of £10 million annual cost savings were made with commitments to reinvest in NBA's brand development. Others speculated that S&N stood to make c. £50 million from the sale of the city centre site to the then Regional Development Agency (RDA) ONE North East, Newcastle City Council and Newcastle University (Walker 2004a) – replicating the property development deal model first tried in S&N's closure of the Fountainbridge Brewery in central Edinburgh in 2004. Despite S&N's production strategy of concentrating investment on fewer, bigger sites in the United Kingdom at Manchester, Reading and Tadcaster (which brewed 5% of NBA's total volume and packaged the brand) none of these breweries had available capacity to take on NBA's volumes at the time (Figure 4.1).

In a locally fortuitous deal, S&N acquired the highly indebted Industrial and Provident Society Northern Clubs Federation Brewery – known as The Fed – across the Tyne in Dunston, Gateshead, for £35.6 million. Disconnecting its historical material geographical associations, S&N relocated NBA production to the newly established company Newcastle Federation Breweries (NFB) and refurbished brewery, shedding 170 jobs by integrating the Tyne and Fed workforces (Tighe 2004) (Figure 4.1). Demonstrating the cultural political economy of considering meaning and value in spatial circuits, the Tyne Brewery closure was not purely an economic decision. S&N's management were torn by the particular geographical associations that originated the history and authenticity in the meaning and value of the NBA brand in the particular 'local' of Newcastle upon Tyne and north east England. Amidst talk of 'Gateshead Brown Ale' and 'Dunston Broon' (BBC Tyne 2004), S&N harboured genuine concerns about the potential damage to the authenticity of NBA's brand meaning and value of shifting its material geographical associations: 'In pure cost terms, it would be cheaper to produce beers from our brewery in Tadcaster, but we are conscious of the scale of our commitment to the North East and also the Tyneside provenance of Newcastle Brown Ale', and were keen to stress it is 'same water supply ... same recipe ... it will taste the same' (Chairman and MD, Scottish Courage, quoted in Walker 2004b: 32).

Using identical specification, brewing parameters and skilled staff, NBA production was transferred to the Fed. The Fed Brewery received a £6 million modernization investment in similar brewing kit to that used at the Tyne Brewery and capacity expansion. In addition, a clandestine 6-month process of 'flavour matching' 'Federation Newcastle Brown' with 'Tyne Newcastle Brown' was undertaken to ensure the output matched the quality and other attributes to avoid arousing suspicions about taste differences.

In terms of the politics of production and geographically uneven development wrapped up in the brand, this local fix 'saved the Fed' (Corporate Affairs Director, author's interview, 2007). Acceptance by trades unions representing members at both sites and anxious to safeguard at least some production and employment on Tyneside stymied public contestation of the Tyne Brewery closure. The lack of social agency contrasted the late 1980s campaign to 'Keep the Broon in the Toon' and 'Don't Give the Dog to the Dingo' – involving public petitions, mass demonstrations and giant banners – mobilized by S&N (then a major Newcastle business employing over 500), Newcastle City Council and trade unions in response to Australian group Elders XL's attempted hostile takeover (see Competition Commission 1989). Despite protests from the Campaign for Real Ale (CAMRA) and the local 'Save the Blue Star' campaign, the response was similarly muted when the Dunston Brewery was closed and put up for

Figure 4.1 Scottish and Newcastle Breweries in Britain, 2009. Source: S&N.
Note: *Closed 2004.

sale in 2009 by S&N's new owners Heineken and NBA production was transferred to the main production site in Tadcaster, Yorkshire – prompting more headlines about 'Tadcaster Brown Ale'. But rather than salvaging any distinctiveness from the Tyne Brewery's architecture that could have provided differentiated meaning and value to any city brand, the site was cleared in haste by its new owners the City Council, RDA and University to make way for the promised glass and steel future of 'Science City'.

In branded commodity stories where beginnings, endings, edges and their delimitations are in question (Cook *et al.* 2006), the geographical associations being used by actors to originate NBA in the particular 'local' of Newcastle upon Tyne and north east England provide a point of entry to understand and interpret the spatial circuits of its meaning and value. NBA's material geographical associations have been severed and relocated but remain inescapable, meaningful and valuable. They shape the agency of the brand's owner and producer at least temporarily in seeking to maintain the production attachment in the territory of the north east – if not the particular 'local' of the city of Newcastle upon Tyne – as part of the brand's differentiated meaning and value in specific market times and spaces. Economic imperatives mediated through the socio-spatial relations of S&N's ownership led to the severing of NBA's material production attachment to the Tyne Brewery and the city of Newcastle upon Tyne, rendering its production spatially mobile and enabling the transfers first to Dunston and then Tadcaster. The manufacture of NBA's cultural meaning and value complicated the unfolding process through the strong material, symbolic and discursive geographical associations constituting its particular 'local' origination, but they were ultimately breakable and shaped the nature of the local production fix. Situating the analytical focus upon the production of the 'local' in NBA by the actors to originate the brand illuminates the material spatial shifts and relations to spatially uneven development; generating geographical differentiation in patterns of output, investment and employment in the economic landscape.

Circulating the 'local' in Newcastle Brown Ale

Echoing the commercial meaning and value of geographical associations in advertising (see, for example, Burgess 1982), the particular 'local' origination of NBA has been central to its geographically differentiated circulation and mobility. Fixing and communicating difference in spatial and temporal market settings (Jackson 2002), the elements of NBA's attributes, facets and cues were cohered by Newcastle Breweries from the outset in the late 1920s by the connection and enhancement of branding into a strong, distinctive and 'unique ... instantly recognizable' brand (Corporate Affairs Director, author's interview, 2007). Early marketing capitalized on NBA's prize-winning recognition at the International Brewer's Exhibition, entered by Newcastle Breweries to raise its profile and test its quality. The 'local' origination of NBA's brand values and equities by Newcastle Breweries drew upon deep symbolic, discursive and visual forms of geographical associations in its socio-spatial origins. This attachment was celebrated at its Golden Jubilee in 1977 as 'Newcastle Brown Ale – born and brewed in Newcastle' (Dobson and Merrington 1977: 6). NBA's particular socio-spatial history forged materially and symbolically strong geographical associations between the 'Brown Ale' and the city (Pearson 1999). The brand's history planted deep roots for NBA in the Geordie public consciousness and culture. Longstanding geographical associations originated in this particular 'local' engendered a strong sense of social ownership and valuation of the brand. S&N recognized this sentiment in that the brand's 'loyal following will claim ... for themselves ... it's our ale, it's our Newcastle Brown ... don't take it away from Newcastle ... it is a powerful force ... and when we do anything with Newcastle Brown we'd be extremely folly to disregard that [sic] ... because it is what's made the beer' (Corporate Affairs Director, author's interview, 2007).

The geographical associations of the 'local' origination of the NBA brand and branding are represented materially, symbolically and discursively in its distinctive packaging. Emphasizing its presentational qualities, NBA has been bottled in clear, flint glass and sized at 550 ml to maintain its equivalence with the traditional pint measure. This differentiated packaging was central to its early advertising (Figure 4.2). On the label, the particular origination and longstanding provenance are connoted by the brewery address, the Newcastle city skyline silhouette of St. Nicholas' Cathedral, the New Castle keep and the Tyne Bridge, and the Blue Star logo – the five points of which represent the founding breweries from Newcastle, North Shields (2), Gateshead and Sunderland consolidated to create Newcastle Breweries in 1890 (Hodgson 2005) (Figure 4.3). Since 1928, the Blue Star has acted as 'a small but very important device adopted to enable the public to distinguish the ales made by Newcastle Breweries without troubling to read the label' (Promotional leaflet cited in Hodgson 2005: 18). Authenticity and cultural authority are represented by the brand name in its original upper case red lettering, 'Brewed since 1927' tag, 1928 commemorative competition plaque, and hops and barley trademarked images. NBA's uniqueness is encapsulated in 'The One and Only' slogan, popularized by Tyneside newspaper *The Newcastle Journal* in 1928 (Newcastle Brown Ale 2007), embossed on the glass bottle (domestic) or on a separate label (export). The secondary label on the bottle's reverse provided discursive space to narrate the 80 years of the brand's particular socio-spatial history. NBA's nickname 'the dog', for example, originated in the gendered vernacular 'I'm gannin' to see a gadgie aboot a dog' used by Geordie men as an excuse to frequent the pub for a drink of NBA (NBA bottle label, 30 November 2006).

Such strong origination in the particular 'local' of the city and the region underpinned the brand's commercial growth. The actors involved constructed NBA's branding to create meaning and value from its 'hardy, working class traditions and values' (S&N 2007: 1) founded in its north eastern urban, old industrial and Geordie context:

> it is a strong brand ...because it has huge identity in terms of the North East ... an icon for the North East ... it has always risen above its peer group ... and we always use the expression 'it punches above its weight' ... if you look at its UK distribution and UK volumes ... it is comparatively small ... but the iconic status of the brand ... is linked to the football club and the city, the shipbuilding ... it is instantly recognisable ... there are some very strong brand cues as well in marketing speak ... clear flint bottle ... the gold band around the label ... the Blue Star and everything ... they are all ... for us very very valuable trademark devices that typify the brand. (Corporate Affairs Director, author's interview, 2007)

Figure 4.2 'Newcastle Champion Brown Ale' advert, c.1928. Source: Historical image, Newcastle Breweries.

The particular 'local' origination of NBA was written through the circulation strategies in its branding, emphasizing its particular socio-spatial history in Newcastle upon Tyne and north east England as well as actively associating and layering the brand in newsworthy 'myths and legends' to generate publicity (Former Marketing Director, author's interview, 2008). The Tyne Brewery in the Newcastle cityscape was signed the 'Home of Newcastle Brown Ale' (Figure 4.4). The 'local' origination of NBA's reputation and folklore were embellished through commemorative bottle labels (e.g. Angel of the North), promotions materials (e.g. posters, banners, umbrellas, inflatables) and brand extension (e.g. recipes, Doddington's ice cream, clothing, memorabilia). Indeed, despite its origination in the city of Newcastle upon Tyne, NBA is considered regionally as a 'north east icon' (Trade Union Official, author's interview, 2007). Sales above the national average in Tyneside, Wearside and Teesside

Figure 4.3 'Newcastle Brown Ale' label. Source: S&N.

transcend the sub-regional rivalries stoked by NBA's historic geographical associations with the city of Newcastle and Newcastle United Football Club through S&N's sponsorship and the NBA label on the club shirt during the 1990s.

From the 1960s, NBA was circulated beyond its north east heartland and became a national brand. It survived rationalization and streamlining within S&N's acquisition-led growth strategy and found its distinctive position as a premium niche brand within S&N's national distribution network. NBA's UK sales volumes peaked in the early 1970s and again in 1990 fuelled by student union cult status and its 'Journey

Figure 4.4 The Tyne Brewery, Newcastle upon Tyne, United Kingdom – 'Home of Newcastle Brown'. Source: Author's image, 2006.

into Space' nickname (Figure 4.5). *NBA*'s differentiated position as a 'local hero' (Corporate Affairs Director, author's interview, 2007) was supported by advertising campaigns. These promotions included 'The One You've Got to Come Back For' based on the pilgrimages of international ethnic stereotypes, including Eskimos and Native Americans, to the Blue Star gates of the Tyne Brewery in Newcastle as well as TV and film product placement (e.g. *Auf Wiedersehen Pet*, *Stormy Monday*) and celebrity association (e.g. Jimmy Nail, *Spender*).

The closure of the Tyne Brewery and relocation of NBA production to Dunston, Gateshead, in 2004 led to questions about whether the brand owner, managers and circulators could still make authentic claims about the brand's origination in circulating NBA when its material geographical associations had changed so fundamentally. Despite stories about the 'end of Brown Ale' (Trade Union Official, author's interview, 2007), the relocation to the Fed was interpreted by S&N as a successful transfer and had 'no effect on saleability' (Corporate Affairs Director, author's interview, 2008). Indeed, the 'local' origination of NBA remained unchanged. NBA's new material geographical association with its principal site of production was referenced only on the bottle label as 'Brewed by Newcastle Federation Breweries Ltd., *Tyne and Wear*, NE11 9JR' rather than the former 'Newcastle Breweries, Gallowgate, *Newcastle Upon*

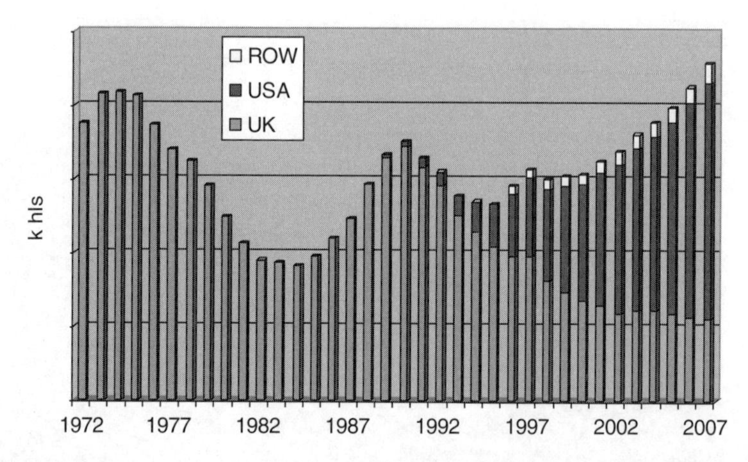

Figure 4.5 Newcastle Brown Ale sales by geographical area, 1972–2007. Source: Jygsaw Brands, Personal Communication, 2008.

Tyne, NE99 1RA'. The particular 'local' origination of NBA acted here as an institutional 'carrier of history' (David 1994) used to communicate the meaning of its sociocultural accumulation of geographically associated brand cues and equities, and retaining its economic value in particular spatial and temporal market contexts where its specific material provenance was ignored or less important. In the context of post-industrialism, the shifting nature of Geordie identity (Nayak 2003) and the production relocation, NBA was considered to have remained 'in the DNA of the city' and still to be internationally recognizable as a 'Newcastle brand' by virtue of the origination of its particular nature and history (Local Government Official, author's interview, 2008). Crucially, NBA possessed such characteristics and attributes whether or not it was still actually brewed and made in the city. In the curve of the Millennium Bridge that replaced the Tyne Bridge on NBA's new bottle label and figured in its new logo (Figure 4.6), NFB in Dunston echoed one of the five founding elements of Newcastle Breweries from Gateshead. In addition, the new brewery's logo fitted neatly the recent cross-Tyne and consumption-oriented rebranding of 'NewcastleGateshead' undertaken by cooperation between the local authorities of Gateshead and Newcastle (Pasquinelli 2014; Richardson 2012).

With a mature and declining ale market in the wake of the shift to lager (so-called 'lagerization'), fragmentation and segmentation into multiple new drinks categories and occasions in the United Kingdom from the late 1980s (Figure 4.5), the logics of capital expansion forced S&N to search for new spatial and temporal market contexts for the NBA brand. The meaning and value of the particular origination in the 'local' of Newcastle upon Tyne and north east England made it difficult to recruit new generations of consumers in the region and beyond as the drinks market evolved. NBA had begun its international travels somewhat earlier, however. Enabled by its bottled packaging and distinctiveness, uniqueness and strong geographical associations, NBA circulated through north east traditions of people working away in response to

(a)

(b)

Figure 4.6 Newcastle Federation Breweries, Dunston, Gateshead, and NewcastleGateshead Initiative. Source: S&N; Newcastle-Gateshead Initiative.

a contracting labour market in the wake of prolonged deindustrialization. NBA was sought as a reminder of home and symbol of belonging: 'Exiled Geordies throughout the world demanded their own brew for special occasions and … the basis for a world-wide export network had been established' (Dobson and Merrington 1977: 6). S&N actively liaised with the armed forces too, providing NBA for naval vessels visiting Newcastle (nicknamed 'Brown Ale Bay') and shipping NBA to regional regiments posted overseas in return for press-worthy photos of 'Newcastle Brown Ale in Hebburn as in Borneo' (Former Marketing Director, author's interview, 2008).

Despite the ambassadorial commercial role of the Geordie diaspora and inventive advertising campaigns such as 'The One and Only world famous British beer' (including straplines 'Hawaii. The Lads' and 'Tyne and every Wear else') touting a national origination of 'Britishness', NBA's international sales outside America translated into premium priced but relatively small volumes across 40 countries worldwide, including China, India and eastern Europe. The particular 'local' origination of NBA's geographical associations was commercially meaningful and valuable only in relatively limited market times and spaces. NBA's value as a branded commercial commodity and its future growth were restored and secured when S&N successfully made one of Daniel Miller's (2002: 27) 'entangled judgements' following its experimental foray into the particular spatial and temporal context of the American market in the early 1980s. S&N both contributed to and caught the wave of premium dark ale niche market growth, new openings of upscale Anglo-Irish bars and interest in imported 'genuine, authentic British … brands' (US importer, author's interview, 2008). S&N's circulation strategy sought to articulate NBA as a 'grass roots level discovery brand' (US importer, author's interview, 2008). Initially, specific entry territories and key trend setting and hip retail outlets were targeted in North and South Carolina and southern California, especially San Diego, rather than the longer established but highly competitive drinking centres such as Boston, Chicago, New York and San Francisco in the largest American state markets.

In this emergent spatial and temporal market setting in America, S&N sought carefully to construct NBA's brand equity based on 'positive product differentiation', distinctive image ('premium, imported, authentic, sense of humour') and taste ('great flavour & easy to drink') (Froggatt 2004). Working a 'subculture affiliation' (Holt 2006b: 373) markedly different from its articulation and circulation in the United Kingdom, the branding of NBA by the brand managers has both constructed and reflected market segmentation of a new niche. NBA's meaning and value has been reoriented to connote and resonate with the consumption cultures of a new, college-educated, typically male and affluent younger generation (aged 25–34) and its independently-styled music, film and travel scene. The producer and distributors' branding efforts have sought to reposition NBA's brand personality as a hip, ironic and 'no-nonsense' ale in a particularly American market context. Crucially, the actors have retained markers of geographical associations constituting its 'local' origination. For example, they have 'kept key brand cues … provided back label stories … about the name and where it comes from … the Star … euphemisms and nicknames … promoting the folklore and authenticity of the brand … its viewed as a brand that's got credentials' (US importer, author's interview, 2008). Branding and circulation activities include advertising its uniqueness (e.g. 'Smooth Like No Other'), humour (e.g. 'If you want bitter don't tip the bartender'), quirky character (e.g. 'The yin and yang of ale') alongside brand imagery of NBA swaddled in 'Old Glory', referencing NBA's Blue Star as 'The 51st Star' (Froggatt 2004), music sponsorship (e.g. Green Day's 2005 American tour), and large-scale and prominent 'brandscape' advertising in selected urban locations (Figure 4.7). The cultural identity formation and status signalling of particular socio-spatial groups has been targeted. Constructions of the consumers as discerning, informed and self-confident drinkers 'fits with trends towards premiumization, trading-up, connoisseurship, badge drinking' (US importer, author's interview, 2008) that S&N seeks directly to connect with to support America-wide distribution and sales.

Cult status has generated sales growth and new nicknames (e.g. 'Give me a Newcastle', 'The Nuke'), supporting brand extension merchandise (e.g. snowboards, clothing and lifestyle accessories, Nike's NBA custom trainers). Even in virtual space, the origination has been carefully handled in on-line interactions. The 'newcastle-brown.com' web site has been targeted at American and international consumers in other growth areas in Australia and China and is distinct in content from the UK 'newcastlebrownale.co.uk' website. Commercially, the reworking of the meaning and value of NBA by its owners, managers and circulators has generated substantial growth. NBA exports to the United States grew dramatically from 30 000 hectolitres (hl) to 441 000 hl between 1991 and 2003, accounting for over 60% of total world-wide volumes of 770 000 hl in 40 countries (Froggatt 2004) (Figure 4.5). By 2005, NBA was the top imported ale in America, against its main competitor Bass, and third ranked grocery 'power brand' (Just Drinks 2006). Against such premium imports, the competition is even mimicking and emphasizing its own 'All American' origination now, such as the Brooklyn Brewery's '*Brooklyn* Brown Ale'. Connecting the cultural benefits of such branding with material economic outcomes, NBA's sales growth in America is highly profitable. NBA six-packs commanded a $1.40 premium relative to its main competitor in the import market and it 'would have struggled to get established without being margin enhancing' (US importer, author's interview, 2008).

Echoing the ways in which 'the performativity of the interface is such that the relation of a brand to an origin may be organised in many different ways' (Lury 2004: 55), central to the owner's efforts to construct and define the NBA brand and its branding in this new spatial and temporal market context has been a shift in its origination. S&N changed the origination of NBA with a specific label marking it as 'Imported from England' in the American market. The brand managers have deliberately associated NBA through material, symbolic and discursive forms of geographical association with a particular 'national' territory within the pluri-national context of the United Kingdom to emphasize its 'imported' and premium market position and address the preferences of its targeted American consumers demanding traditional 'English' ale. The particular 'local' origination associating NBA with the city of Newcastle and north east England central to the socio-spatial history of the brand and its branding has not been played up by the actors involved. This tactic has been used because 'in the US NBA's main competitors Bass, Guinness are really known as imports from the UK and [we] didn't feel we could make a gain by making a big deal' (US importer, author's interview, 2008). Moreover, S&N's importers and marketers recognize that:

> there is much less cogniscence amongst US consumers about where it might be brewed and much less expectation that it would all come from a place called Newcastle … it can send mixed messages … it's obviously changed and is coming up in the world … but in other ways it can be interpreted as a hard drinking beer from a hard working town … it's not necessarily a position we would want to take. (US importer, author's interview, 2008)

S&N's efforts in originating NBA's geographical associations as 'Imported from England' distanciated the brand from the Tyne Brewery closure and Dunston and Tadcaster relocations. This strategy worked as a means of cohering meaning and value especially given the lack of knowledge in this specific spatial and temporal market context: 'the Americans couldn't give a shit if it was brewed in Sunderland, Gateshead, wherever it is … they want English beer, they want Brown Ale … they want a dark beer and as long as it comes from England they're happy' (Trade Union Official, author's interview, 2007).

As brand owner and producer, S&N and subsequently Heineken now face a dilemma in organizing NBA's spatial circuit to enable and underpin the particular 'Imported from England' origination central to its meaning and value. Growth in NBA's sales volumes has made brewing in America economic especially as brewing is a classic weight-gaining industry in Weberian terms in which scale economies and proximity to market are critical. But producing the brand in America may be commercially damaging for the authentic meaning and value of the 'Imported from England' origination and label:

> it's economic now … it should be produced and packaged in America … we've done trials … where we have actually brewed it here and sent it over to America to be packaged … but the marketing people are very sensitive to that … that's the only reason that we haven't gone ahead with it … if the American market found out that it was packaged in America it would affect that market … but there are some huge distribution savings to be made … because it's a heritage brand … our sales people would get nervous about it … as long as you can charge a premium price for it … it is a very profitable brand for us. (Brewery Manager, author's interview, 2007)

The meaning and value of the 'Imported' category in this particular spatial and temporal market context even prompted S&N to look further afield to source NBA: 'we wanted to retain the "Imported from England" label … this provides cachet for the brand … it must have the brand message … but we looked at whether that should just be 'Imported' to allow us to brew in Mexico or Canada' (US importer, author's interview, 2008). Substance to the rumours of 'Tijuana' or 'Toronto Brown Ale' has yet to materialize.

Reflecting the mutability of branded commodities over time and space through branding (Smith and Bridge 2003), geographical associations and the origination in the bounded territory of Newcastle upon Tyne and north east England were constructed and cohered by the brand owner, managers and circulators in the NBA

Figure 4.7 Newcastle Brown Ale 'brandscape' advertisement, United States, 2006. Source: S&N.

brand in material, symbolic, discursive and visual forms. The rich geographical associations of a particular urban and regional social history were used to originate NBA in the particular 'local' of Newcastle upon Tyne and initially shaped its circulation in new markets beyond the north east in the United Kingdom. Following the Tyne Brewery's closure in 2004, NBA's relocation to Dunston and then Tadcaster were not highlighted in its origination and circulation. The perceived origination of the brand cohering its meaning and value was undisturbed and even blurred by the new cross-territorial 'NewcastleGateshead' place brand. Building upon its longstanding internationalization and the economic imperative to secure new and growing markets, the brand owner's importing business in America focused on entering a particular spatial and temporal market context but had to subtly to undertake the cultural reworking of its 'mystical veil' to utilize the bulk of NBA's brand equities to connect to a particular and growing sales demographic. Geographical associations to an identifiable and authentic origin place resonated across time and space as the circulation of NBA's value and meaning was instrumentally reoriented by its circulators from origination rooted in the 'local' of the city of Newcastle upon Tyne and north east England in the United Kingdom to (sub-)national origination framed as 'Imported from England' in America. This changed origination demonstrated the tensions and accommodations between bounded, territorial spaces at different scales and their mobility and projection in unbounded, relational spaces. The success of NBA's circulation strategy in its new-found cultural resonance and market value then raised issues for its brand owner about the location of production within its spatial circuit of value and meaning.

Consuming the 'local' in Newcastle Brown Ale

Interrelating its differentiated brand equities, technical characteristics and packaging, NBA's particular 'local' origination is strongly geographically associated with distinctive consumption practices resonant of its socio-spatial history in north east England. Traditionally, the 550 ml bottle is decanted into a half-pint Schooner or Wellington glass to maximize its taste and sustain the head. This locally particular 'style of drinking' was critical in 'perpetuating the myth' of NBA's unique aesthetic and sensory appeal:

> the design of the bottle was vital because it had this high shoulder … this immediately recognizable shape … in the mechanics of pouring the beer it actually provided this glug, glug, glug … which produced the head, an effervescence in the thing, and the sparkle came out and … the nose came out, the smell … keep refreshing the glass … it is the only way to drink it … because every glug bursts the thing up again … brings out all its qualities … its visual appearance was enhanced enormously by going through this ritual. (Former Marketing Director, author's interview, 2008)

Such regionally rooted consumption practices reinforced the meaning and value of NBA's distinctive brand identity. Sales grew beyond its north east heartland to its early 1970s peak and decline, triggering wider distribution to south east England and the Midlands and its resurgence during the 1980s prior to the fall and stabilization in the 2000s (Figure 4.5).

In the UK, NBA has dominated a static and at times contracting market as a differentiated brand capable of sustaining a premium price in a commercial context of less loyal consumers with widening repertoires of brands and drinking occasions. In the on-trade bottled ale segment in bars, pubs and restaurants, NBA grew its market share to a peak of 42% in 2003, against the much more locally distributed Manns (9.3%) and Whitbread Pale Ale (7.2%) (S&N 2004). In a market niche in the United Kingdom in which heritage and provenance are increasingly valued by consumers, NBA has sustained its strong market position, despite limited marketing investment relative to S&N UK's 'drive brands' and its production relocation to Newcastle Federation Breweries in Dunston, Gateshead, and then Tadcaster Yorkshire. Indeed, NBA is even seen by some consumption actors as 'a NewcastleGateshead brand now' (Place Marketing Chief Executive, author's interview, 2007). It is situated within the 'NewcastleGateshead' place brand, presented internationally at trade fairs as a resonant marker of a particular local identity, used in destination marketing as an internationally identifiable shorthand for the place, and presented to visiting dignitaries with bespoke commemorative labels: 'Newcastle Brown is a great selling story for people coming into the area and there is a bit of actually being brewed in Gateshead in some ways "de-Newcastle-izes" it ... brings it more part of the north east rather than just being a small part of the north east' (Local Government Official, author's interview, 2008).

In a contracting UK market setting for ale from the 1990s, the geographical associations used by the brand managers and circulators in their attempts to originate NBA in a distinctive and unique 'local' to establish and articulate its brand identity faced constraints on their growth. As the spatial and temporal market context shifted, consumption mores created a cachet around the brand unhelpful to S&N's attempts to attract replacement generations of younger, especially women, drinkers. Acutely for NBA, its particular socio-spatial and class histories were an 'important part of the culture' of 'Brown Ale, flat caps, whippets, leek growers' but were pitched into tension with the new place brand's 'iconography of the region ... the Angel of the North ... Sage Gateshead ... life sciences ... blending heritage and innovation' (Place Marketing Chief Executive, author's interview, 2007). The brand management's attempts to mimic and adapt NBA's circulation strategies from the American market – including a bottle label designed by Newcastle band Maxïmo Park and 355 ml bottle drinks promotions in Newcastle's Digital night club – have 'failed to take off' (Former Marketing Director, author's interview, 2008). Particular and traditional regional consumption norms have endured and troubled attempts at change: 'Halves of Brown Ale? Never heard of it ... especially in this region' (Trade Union Official, author's interview, 2007). Indeed, effectively admitting the non-core position and relative neglect of its 'heritage brands' such as NBA just prior to its takeover by Heineken, S&N sold NBA's sales and marketing rights in the United Kingdom to newly established spin-off company Jygsaw Brands. Their strategy is rediscovery of 'the Geordieness of *NBA*' (Sales and Marketing Director, author's interview, 2008) and reconnection of the brand with its core heartland, reawakening awareness through the 'I ⋆ NE' campaign, 'Broon Tour' city map and construction of 'brandspaces' in key outlets in the city with 'point of sale in every bar in Newcastle that is shouting Newcastle Brown to the people that walk in ... then you start to engage this "The city feels like a Newcastle Brown city again"' (Sales and Marketing Director, author's interview, 2008).

In its circulation and sales growth in America since the early 1990s, consuming the geographical associations constituting the 'local' origination of NBA is markedly different from the United Kingdom. Consumption practices have combined specific characteristics of the particular spatial and temporal market context in America. This has included the shift in demand from commodified lagers towards dark and imported beers and the taste for cold, carbonated and sweet, flavoursome drinks. These particular consumer characteristics have connected with NBA's specific attributes, especially the changed origination of 'Imported from England', marked through its labelling and establishing its provenance. Drinking NBA in America is mostly from draught in pint glasses (75% of on-trade sales) or 355 ml bottles (25% of on-trade sales) and cold, either dispensed (3°C compared to 6°C in the United Kingdom) or from the bottle straight from the fridge. In competition with other imports Bass and Guinness and regional micro-/craft beers (e.g. Sam Adams, Boston; Fat Tire, Colorado), NBA provides for its target market segment a 'story in relation to other darks ... [an] unconventional choice for consumers because of the flavour profile of the beer ... consumers enjoyed the personality of the brand' (US importer, author's interview, 2008). The meaning-making of NBA's particular consumers creates and sustains the brand's premium priced market value and high margin. Brand owner S&N has reaped substantial commercial benefit from this new and growing market segment and the changed origination of the brand. NBA is 'probably the best example in our group of how one brand has been transformed from one part of the world to another' (Head of Corporate Communications, S&N, quoted in Whitten 2007: 1).

The impact of the relocated material geographical association of production from Newcastle to Gateshead is interpreted as negligible in this distanciated spatial and temporal market setting in which NBA's origination has been subtly shifted to 'Imported from England':

> [the relocation] makes no difference ... [the] move to Gateshead was only important locally. ... Beyond 10 miles of the city does anybody notice or care? To all intents and purposes it is brewed in Newcastle ... if you're in a bar in Vietnam, in LA ... it's brewed in Newcastle ... internationally it's still Newcastle Brown Ale. (Corporate Affairs Director, author's interview, 2007)

Indeed, the sales growth in America provided a marked increase in demand and raised capacity constraints for NBA production at Dunston leading to its closure and further relocation to Tadcaster, Yorkshire. Echoing Ray Hudson's (2005) more open and porous notion of virtual 'spaces of brands', the growing consumption of NBA in America has extended its geographical reach in relational space through web-facilitated brand extension merchandizing and interactivity including competitions, downloads and promotions. Even in on-line relational space, NBA's international web site content is spatially targeted, reflecting the continued differentiation of the brand's geographical associations and origination in its circulation and consumption in particular market times and spaces.

The geographical associations constituting the particular originations of NBA have been used by the participant actors involved in attempts to construct and cohere the meaning and value underpinning its consumption in territorial and bounded spaces and places. The agency of not just brand owners and producers but the circulators of advertisers, importers, marketing consultants as well as consumers in the United Kingdom and internationally and the regulators of local authorities and place marketing institutions have each played their different roles within the brand's spatial circuit. Blending material attributes and culturally constructed ritual, consuming the 'local' origination of NBA in north east England involved particular practices that added layers to the brand's 'mystical veil' and reinforced its distinctive differentiation. NBA's meaningful and strong consumption identity maintained its value in a shifting market context during the 1990s, despite the relocation in its material geographical associations of production in the context of the emergent and branded entity of 'NewcastleGateshead'. But as the UK ale market declined such locally originated practices jarred with new generation consumers and compelled the brand owner's search for new sales and marketing opportunities in America. NBA was successfully connected by S&N with the markedly different consumption culture and practices in a different market time and space in America in the 1990s. Framed by its circulatory representation, despite its shifting spatial attachments in Tyneside, the changed (sub-) national origination of 'Imported from England' prefigured its meaning and value amongst this specific clientele.

Regulating the 'local' in Newcastle Brown Ale

For all their cultural construction of meaning, in the brewing business *NBA*'s owner saw:

> brands as vehicles for driving growth and value – no more and no less. What we're looking at is very unemotional. We can talk about the beauties of brands and all that sort of stuff, but at the end of the day investors just want to know what they are going to get in terms of returns. (Tony Froggatt, S&N, Chief Executive quoted in Bowers 2006: 30)

As the meaning and value of branded goods and services commodities are inescapably geographically associated and integral to their origination, S&N own and control these spatial connections and connotations wedded to the brand through rights to the brand name, 13 trademarks – 'all of the devices that make that Brown Ale' (Corporate Affairs Director, author's interview, 2007) including the combination of bottle, label and Blue Star – and commercial confidentialities (e.g. the recipe, production specification). Sometimes neglected in accounts of brands and branding (except, for example, Da Silva Lopes and Duguid 2010; Lury 2004; Lury and Moor 2010), ownership and valuation of these assets is critical in spatial circuits as regulatory 'tools of intellectual property' that 'are at once discursive and material: they function as a kind of holding operation on the flux of social life in order to identify commodity authors and name the recipients of surplus value at given geographical scales' (Castree 2001: 1523).

In addition to the regulation of the NBA brand as a financial asset owned by S&N plc, the geographical associations central to the origination of NBA in the United Kingdom and European Union (EU) markets were formalized through the EU's Protected Geographical Indication (PGI) scheme. As S&N pursued its internationalization strategy it sought further brand differentiation for NBA that would construct meaning and value in the particular spatial and temporal market context of the EU during the 2000s. S&N too sought to reinforce its control and protection against copyright infringement as it entered new and emergent markets in central and eastern Europe. S&N successfully applied for EU PGI status for NBA in 1996 based on its 'unique recipe ... developed to meet local tastes using yeast grown at the brewery and a unique salt/water mix' (Department of Environment, Food and Rural Affairs (DEFRA) 2006: 2):

> the PGI ... was seen as ... one of the ways we could fairly swiftly throw some protection around the brand within ... the European Community ... it was also seen as a very good marketing tool ... having the cachet of an identified geographic region puts you a cut above any of your competitor brands ... [a] springboard into Europe in terms of marketing to say ... unique to the north east of England ... you can't copy it ... you can't duplicate it ... and the PGI documentation specifies a unique brown ale brewed according to a recipe owned by S&N at the Tyne Brewery, Newcastle ... it tied it to a specific recipe, it tied it to a name, it tied it to a site. (Corporate Affairs Director, author's interview, 2007)

In line with the PGI scheme's aims of recognizing and protecting the provenance and value of especially agro-food commodities (Parrott *et al.* 2002), S&N's application claimed that 'Water is taken exclusively from the area' and 'the added yeast and salt/water blend are unique to the Tyne Brewery' but 'All the remaining ingredients are sourced from the UK' (DEFRA 2006: 1). In addition, S&N stressed that the NBA brand's 'trade secrets' (e.g. ingredients, processes, recipe) were trademarked, owned and controlled by the brand owner. This specific articulation raised several concerns that appeared to undermine NBA's claim to PGI status under the ethos and terms of the scheme. First, the specification of particular local *and* extra-local geographical associations suggested some ambiguity in the degree to which NBA's intrinsic and specific characteristics were attributable solely to its production location in a spatially delimited area on Tyneside. Indeed, NBA did not qualify for the EU's Protected Designation of Origin (PDO) with its tighter regulation of provenance and origination. Second, PGIs were usually awarded to consortia or groups of producers located in specifically defined geographical areas that own the brand name (e.g. Parma ham) rather than a single company that owns the brand and produces it at a single site. Although established in 1927 and 'inextricably linked to Newcastle ... there weren't necessarily specific characteristics of the product that were due to its geographical location' (Central Government Official, author's interview, 2006). Echoing interpretations of PGI as a local development tool for embedding and attaching assets in place (Morgan *et al.* 2006), rumours circulated that local management applied for the PGI as a means of protecting the Tyne Brewery's future by tying it to the NBA brand only in Newcastle upon Tyne. S&N's decision to close the Tyne Brewery and relocate production beyond the territory specified in the PGI undermined the basis of NBA's PGI. This was especially the case because S&N were unwilling to redraw the specified geographical boundaries of the PGI territory across the Tyne to Dunston because it might be a 'millstone if they might want flexibility to move production' (Central Government Official, author's interview, 2006). This stance proved prescient as Dunston was subsequently closed in 2010 and NBA production relocated to Tadcaster.

In 2004, S&N concluded the PGI could not technically prevent them closing the Tyne Brewery and prefigured their unprecedented step of applying to the EU to revoke NBA's PGI:

> production at the site in the Newcastle-upon-Tyne city ... is no longer commercially viable ... [and] presents operational difficulties ... [S&N is] to close the Newcastle Brewery and to move to another site in the north east of England. Therefore the specification is not any longer respected in relation to the delimited geographical area of the PGI that is the city of Newcastle-upon-Tyne. (S&N application to revoke the PGI, European Commission 2006)

Mandatory notification in the Official Journal of the EU allowed public contestation. CAMRA argued that the PGI should be upheld. The rationale was that even if S&N no longer wished to produce NBA within the specified area other producers might and if the PGI was cancelled then the currently regulated geographical and historical association between NBA and Newcastle central to its particular locally originated meaning and value would be severed, opening up the potential for further production relocation. For CAMRA:

> Consumers would be misled as [NBA] is intrinsically linked to the city of Newcastle Upon Tyne in the minds of consumers ... [S&N] state that the added yeast and salt/water are unique to the Tyne Brewery ... allowing producers to apply for the cancellation of a registration so that they can shift production in order to maximise profits undermines the credibility of the EU protected names scheme. (CAMRA Letter to DEFRA, 7 June 2005)

Despite CAMRA's challenge, the UK government ministry responsible at DEFRA recommended the revocation of NBA's PGI to the EU because of its peculiar nature. The PGI scheme aims to offer protection for consortia of multiple producers of collectively owned products meeting precise specifications in a defined geographical area (Morgan *et al.* 2006). In contrast, NBA was a brand owned and trademarked by a single producer (S&N), a publicly limited company and registered for a PGI at a single production site (the Tyne Brewery). S&N's 'commercial logic' argument was decisive because 'the only person that can challenge against this would be a fellow producer of Newcastle Brown Ale ... by definition there are none ... any strong trademark protection effectively trumps the PGI thing' (Corporate Affairs Director, author's interview, 2007).

S&N's approach proved controversial with local political representatives arguing that:

> Recognizing that it never was a soundly based PGI within European rules ... the PGI status should never have been granted ... it was always really a brand ... and its legal

> status as a brand was always under the control of the company ... the European Commission ... decided that fundamentally the PGI had no independent existence beyond the intellectual property of the brand and therefore could be surrendered. (Member of Parliament, author's interview, 2008)

Any potential new market entrants seeking to produce a branded commodity named anything like NBA would infringe the copyright enshrined in NBA's trademark protection by S&N, not be able to gain access to S&N's commercial confidentialities of recipe and production specification, and face substantial barriers to market entry relating to scale, distribution and marketing. In the wake of the PGI episode, subversion and homage to the NBA brand prompted local microbreweries in Tyneside and the wider north east region to produce Geordie Pride's 'Toon Ale', Hadrian and Border's 'Byker Brown' and Mordue's 'Wallsend Brown Ale'.

No longer encumbered by formally regulated geographical associations to the Tyne Brewery in Newcastle, coupled with NBA's sales growth in America, S&N's new found spatial flexibility and new owner Heineken's international capacity stoked anxiety about the further mobility of the NBA brand's production: 'Technically, we could go anywhere in the world to produce Newcastle Brown. ... If we can go to Dunston, we can go to the Moon' (Company Spokesman, S&N, quoted in Whitfield 2006: 2). Concerns emerged locally that 'they could sever their links with Newcastle even further. It could be Ohio Brown Ale just as easily' (CAMRA Representative quoted in Whitfield 2006: 1). S&N's incorporation into new owner Heineken's strategy and structure of spinning-off 'Heritage Brands' and moving towards increased contract brewing even raised the prospect of selling-off the NBA brand:

> what they would do is sell the Brown Ale brand ... say they sold the Brown Ale brand to Joe Bloggs in America he would then decide what is the most economical way to get it manufactured ... he wouldn't necessarily say 'I've bought the Brown Ale brand and that also gives me a brewery in Dunston' ... it would be 'I've bought the Brown Ale brand now I need to gan [sic] and market it' ... wherever in the world ... and I'll buy it from where I can get it ... I mean all consumer brands are like that now. (Brewery Manager, author's interview, 2007)

While the wholesale spin-off of NBA has yet to happen, the further mobility of its production has occurred following the closure of the Dunston brewery, Gateshead, and relocation of production to Tadcaster, Yorkshire. The geographical associations integral to the particular 'local' origination at the heart of NBA's meaning and value have been (re)constructed, regulated, disrupted and contested by a range of interrelated actors in its spatial circuits: the brand owner and managers; the national government department; the European Commission Directorate; the local Member of Parliament; and, the campaigning organization. Tensions were evident between the bounded, territorial regulation of the PGI that tied the production of the NBA brand to a specified and regulated territory, establishing and representing its 'local' origination, and the unbounded, relational agency of the brand owner S&N enabled by its ownership and control of NBA's trademarks and commercial secrets. CAMRA sought explicitly to unveil the regulated form of the 'commodity fetish' in the PGI and retain its tie to its specified production site of the Tyne Brewery and territory of Newcastle upon Tyne and the north east region. But S&N's power of brand ownership reproduced geographically uneven development in decisively exercising its spatial flexibility in relocating its material geographical associations of production from its historical origin to Dunston and later Tadcaster.

Summary and conclusions

Uncovering the particular socio-spatial history and geographical associations of NBA have been integral to explaining how its particular 'local' origination proved powerful to the brand's owners and managers in creating and cohering meaning and value in its spatial circuits. But in the specific spatial and temporal market setting of the United Kingdom in the 1990s, NBA's owners had to grapple with its origination and strong attachments to a particular declining market segment and region in north east England. Elsewhere in America, NBA's owners, importers and marketers subtly changed the origination of the brand from its 'local' geographical associations only to Newcastle upon Tyne in north east England to the (sub-)nationally framed 'Imported from England' to tap into a new and rapidly expanding market. The different ways in which the origination of the brand matters (or not) in different spatial and temporal market contexts shapes territorial development through influencing decisions about where economic activities take place and investment, supply contracts, jobs and so forth are located. In the United Kingdom, the brand owner's 'local' origination of NBA has not prevented the severing of the brand's material geographical associations of production with its historic origins following the closure of the Tyne and later Dunston Breweries, with the loss of over 300 jobs, and the relocation of NBA production to Tadcaster in Yorkshire. NBA's changed origination as 'Imported from England' in the American market is posing very different questions about what impact it would have on its marketing if it were more economically brewed in America or produced and originated as 'Imported' from Canada or Mexico. NBA's socio-spatial biography reveals how origination of branded commodities by the actors involved are not fixed or permanent constellations of geographical associations creating and fixing meaning and value; they are constantly disrupted and shaped by the agency of actors in coping with the rationales of accumulation, competition, differentiation and innovation in particular market times and spaces.

Chapter Five
'National' Origination ... Burberry

Introduction

The actors involved in the brand and branding of Burberry have sought to construct a 'national' origination, evoking a particular version of the nationally framed and rooted geographical imaginary of 'Britishness' in attempts to create and cohere meaning and value in the spatial and temporal market contexts of the fashion business internationally. The particular socio-spatial history of Burberry has afforded its brand owner, managers, marketers, shoppers and regulators pliable sources of discursive, material and symbolic geographical associations that have enabled efforts to construct meaning and value based upon distinctive attributes of authenticity, quality and tradition. Origination reveals how the meaning and value of the particular version of Britishness deployed that propelled Burberry's steady post-war growth was superseded by economic, social and cultural shifts in its spatial and temporal market settings. The brand was left exposed and reliant upon a narrow array of core products and limited, conservative consumer base. Burberry's adverse commercial prospects were worsened by uncontrolled international brand extension, distribution and licensing. The new management's revitalization of the origination of the brand's Britishness has been integral to its commercially successful modernization. The heritage assets and their geographical associations have been reworked to update Burberry's image and market positioning, internal control over manufacturing and distribution has been increased, and the product portfolio has been selectively expanded. Yet internationalization, ongoing transitions in market times and spaces, and subcultural appropriation have further disrupted the nationally framed geographical associations utilized by the

Origination: The Geographies of Brands and Branding, First Edition. Andy Pike.
© 2015 John Wiley & Sons, Ltd. Published 2015 by John Wiley & Sons, Ltd.

actors involved in trying to originate Burberry's meaning and value in the geographical imaginary of Britishness. Pressures for future growth and financial returns transmitted through the Burberry brand's publicly limited company ownership structure have encouraged the internationalization of the brand's distribution and supply base as well as its rationalization, resulting in factory closures in the United Kingdom and mobilizing the 'Keep Burberry British Campaign' led by local actors. Origination deepens the understanding of the brand's predicament by qualifying any economically determinist explanation. This analytical frame demonstrates that even as the brand's material geographies of production are internationalized *beyond* the specific national territory of Britain, the creative design, styling, detailing and advertising remain originated *in* and *with* Britain as integral constituents of its meaning and value within its spatial circuits. This origination is produced, circulated, consumed and regulated in particular spatial and temporal market contexts internationally seemingly independent of the brand's material geographical associations of production in specific locations.

Producing Britishness in Burberry

Established by Thomas Burberry in Hampshire, England, in 1856, the origins of Burberry stretch back more than 150 years. Central to Burberry's socio-spatial history and survival has been the brand's intertwined relationship with the meaning and value of particular versions of Britishness. As a nationally framed set of geographical associations that has undergone constant reinvention and defies specific or fixed definition, Britishness is a powerful and longstanding marker of authenticity, quality and tradition in fashion brands and branding internationally (Goodrum 2005; McDermott 2002; McRobbie 1998). As its enduring global fashion centre, the city of London in the United Kingdom is inextricably intertwined with Britishness (Breward and Gilbert 2006). Britishness provides an abundant and malleable source of valuable and meaningful assets and traditions for actors involved in fashion brands and branding. Specifically, the Burberry brand's socio-spatial history is replete with connections and symbols of Britishness, supplying the state as a military outfitter and the monarchy as a Royal Warrants holder (Burberry 2005). In the Burberry brand archive, these resources exist in discursive, material and symbolic forms including advertising copy, images and innovations such as the 'breathable, weatherproof and tearproof' gabardine fabric developed for British military use in the age of Empire (Burberry 2006: 2). Such sources of meaning and value constitute the 'power of the Burberry brand' (Angela Ahrendts, Chief Executive Officer, quoted in Burberry 2007: 41). These assets are codified as its 'core values' manifest in its 'brand icons' of the Prorsum ('forwards' in Latin) horse logo, trenchcoat and signature check (Burberry 2005: 86) (Figure 5.1).

Regaining control over the Burberry brand and the origination of a particular version of Britishness in its branding have been integral to the management's successful turnaround and corporate growth into a £1 billion sales revenue company under the leadership of Chief Executive Rose Marie Bravo from the late 1990s. As a

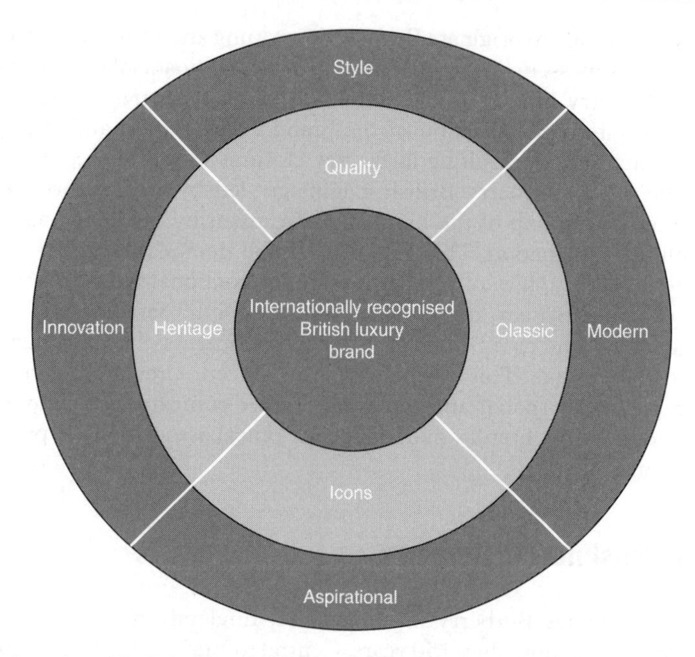

Figure 5.1 Core values of the Burberry brand. Source: Adapted from Burberry (2005: 86).

'neglected backwater of the GUS [Great Universal Stores] empire' (*The Economist* 2001: 1), the Burberry brand suffered underinvestment in core functions including product development, marketing, merchandising and other support infrastructure. With a decentralized organization and holding licensing relationships with broadly based retail groups selling multiple brands in key markets (e.g. Mitsui, Japan), the brand had been loosely controlled within the wider GUS corporate structure, fostering multiple loci of design and branding.

From the late 1990s, the new management's strategy was brand-led. It was based upon updating the Burberry brand image and multi-brand positioning, increasing internal control over manufacturing and distribution, and more carefully expanding the product portfolio across a wider customer base (Moore and Birtwistle 2004). Following the fashion trend of appointing high profile 'star' designers situated within the networked infrastructures of the 'key sites' of global fashion capitals (Weller 2007: 39), brand design for Burberry was centralized in an enlarged, consolidated and international team in 2001 under the control of a single Creative Director, Christopher Bailey (formerly of the international fashion brands Gucci and Donna Karan):

> Burberry centralised its design process in London, largely eliminating regional design centres, and physically integrated the category specific design teams, to create a single, integrated design function. As part of enhancing the clarity of in-store presentations, product teams reduced significantly the number of styles developed within each collection. The product design cycle was revamped to increase the number of collections

> created, allowing new merchandise to be offered in stores monthly. These design and merchandising initiatives have been enabled by increased investment in product development and design and merchandising talent. (Burberry 2007: 44)

The Creative Director drew upon the 'the DNA of the new Burberry' (Rose Marie Bravo quoted in Schiro 1999: 1) in the historical 'combination of stories, archives and legitimacy' (Fashion Industry Analyst, author's interview, 2008) of the brand. The actors sought to rework and originate Burberry's distinctively 'British' heritage assets for the new Millennium. New versions and twists were introduced to capture and anticipate the zeitgeist and produce a reinvigorated 'cool' and younger brand (Tungate 2005). As part of the Burberry Group's new £22 million headquarters in London, the new design centre was connected and informed by outposts providing input from key markets worldwide, especially Japan and Spain, to ensure the consistent and coherent origination of the geographical associations in the brand and the fine-tuning of their meaning and value to capitalize upon the attributes in specific spatial and temporal market settings.

In trying to reconstruct and fix meaning and value in the Burberry brand, a redefined and fuller brand pyramid was established to provide a brand hierarchy to guide group and branding strategy and production priorities to 'align sourcing origin with product pyramid' (Burberry 2008: 46) (Figure 5.2). The aim was 'broad coverage of product categories and differential price positioning among the brands' to provide 'a comprehensive lifestyle offer that also enables customers to access, as well as trade-up (and down) between the various brand levels' (Moore and Birtwistle 2004: 416). The premium brand Prorsum collection occupies the apex through its 'luxurious' and 'fashion forward' lines that provide 'design inspiration' (e.g. colour palettes, silhouettes) for the brands beneath (Burberry 2008: 51). The London Collection constitutes the tailored ready-to-wear middle section of the pyramid and is differentiated by the 'core international' and historically 'localized' (Japan and Spain) collections. The base of the pyramid comprises the Lifestyle diffusion brand collections with entry-level casual sportswear products to introduce and encourage brand awareness, literacy and loyalty.

The Burberry brand hierarchy and its geographical associations with a particular version of the nationally framed origination of Britishness across different product ranges explain why the story of fashion production is not just an economically deterministic one of inevitable international outsourcing to relatively lower cost locations and/or upgraded suppliers (cf. Jones and Hayes 2004). Scrutiny of how the geographical associations of production are organized by the Burberry production managers illustrates the spatial diversity of 'make or buy' decisions (Pickles and Smith 2011). Variety is generated by the geographical associations with the origination of Britishness central to Burberry's meaning and value and the connotations and perceptions of how this Britishness is produced in relation to the specific national territory of Britain. For each particular branded commodity, production strategy is determined by numerous and related factors. These include the 'iconic sense of the Burberry brand' and 'quality of manufacture' (John Peace, Chairman, Burberry quoted in Welsh Affairs Select Committee 2007: 4), craftsmanship, materials, style and reference to the brand hierarchy together with the economic imperatives of market analysis, desired price point, unit

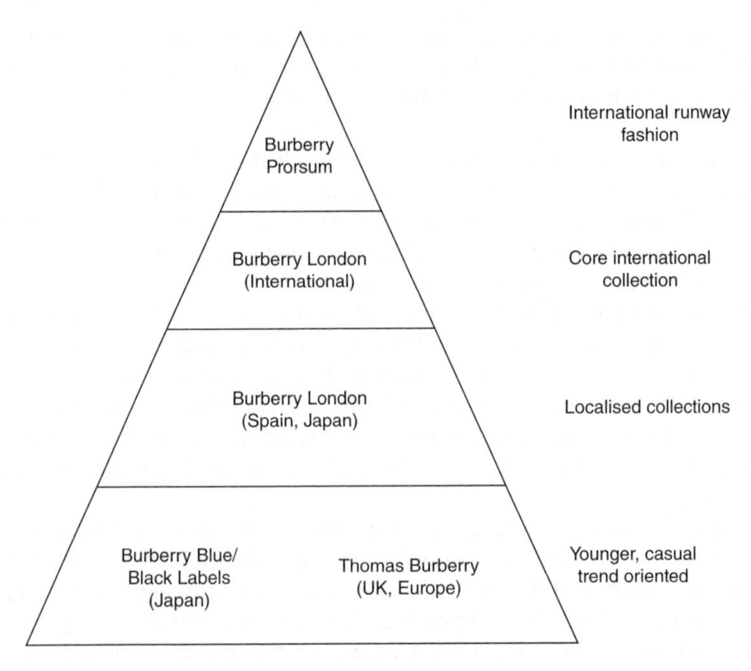

Figure 5.2 The Burberry brand pyramid. Source: Adapted from Burberry (2005: 86).

cost, volume, and 'continued style reduction to improve sourcing efficiency, order fulfilment and retail productivity' (Burberry 2008: 44).

Differentiated by specific goods, Burberry remains to a degree a vertically integrated 'retailer with factories'. In-house manufacture is reserved for signature items in Burberry's product range where the authenticity, tradition and high quality of being originated as the nationally framed 'Made in Britain' are integral to meaning, value and premium price. The brand's characteristic rainwear and the 'brand icon' trench coat are produced in Britain in wholly owned factories in Castleford and, until recently, Rotherham, South Yorkshire (Figure 5.3). Significantly, and demonstrating the relational construction of Britishness in connection to Burberry's perceived British roots and heritage, much of the premium brand Prorsum collection is produced in Italy (Tokatli 2012a). Burberry exercises close relations and control in the procurement of other key inputs where high quality, finishing and provenance are integral to the brand's meaning and value. This sourcing includes textiles for linings and fabrics from wholly owned subsidiary Woodrow-Universal, Keighley, West Yorkshire, and leather and knitwear from long-term suppliers elsewhere in Britain and in continental Europe, primarily in Italy (Burberry 2009). Indeed, as the supplier base has been rationalized from 240 to 100, sourcing strategy has shifted from 'cut, make and trim' to 'full package' relationships with fewer and more capable suppliers located within Britain and, increasingly, internationally through intermediaries in Italy, Turkey and Hong Kong (Tokatli 2012a).

Figure 5.3 Burberry operations in Britain, 2010[a]. Source: Based upon Burberry (2005, 2009).
Note: [a]Excludes three additional United Kingdom sites (wholesale warehouses, locations undisclosed).

Lower down the brand hierarchy amongst the lower value-added and commodified items less 'recognised as being made by a British brand, Burberry' (John Peace, Chairman, Burberry quoted in Welsh Affairs Select Committee 2007: 4), cost pressures are more keenly felt and price points, market positioning and the competition are more carefully scrutinized. Being originated as 'Made in Britain' is less important to consumers of these branded goods in their particular spatial and temporal market settings. These more peripheral products are typically outsourced internationally to lower cost subcontractors in eastern Europe, Morocco, Turkey and, latterly, China. Supporting the brand's extension and recent forays into 'co-branding', Burberry utilizes specialist branded global supply partners for particular hard and soft non-apparel items, including eyewear (e.g. Luxottica, Milan), fragrance (e.g. InterParfums, New York) and timepieces (e.g. Fossil, Richardson, Texas), often replacing local suppliers.

The tensions and accommodations involved in producing the material *and* perceived geographical associations of Britishness central to the origination of the Burberry brand erupted in 2006. As part of an overhaul of corporate systems and its supply chain, Burberry announced the closure of its factory in Treorchy, Wales, with the loss of 309 jobs. The factory was deemed 'no longer commercially viable' and production was planned to be relocated to existing suppliers that already produced 75% of Burberry's polo shirts in southern and eastern Europe and a new supplier in China (Burberry Spokesperson quoted in *Western Mail* 2006: 1). Polo shirts constituted a branded product from the lowest and most cost sensitive segment of the brand pyramid. Local campaigners including the community, Burberry workers, political representatives and trade unions targeted the brand in their efforts to prevent the closure. The campaign enrolled Welsh and other celebrities in their bid to 'Keep Burberry British' (Figure 5.4), questioning the brand's claims to the value and meaning of the origination of Britishness in the wake of its international outsourcing. Campaigners questioned whether it was in breach of the Royal Warrants granted to Burberry as a supplier to members of the monarchy by making the brand outside Britain and potentially utilizing sweatshop labour. The discontinuity of the geographical associations at the heart of the brand's meaning and value was integral to the sense of injustice: 'the hypocrisy of a multinational brand whose global marketing campaign revolves around "Britishness" and being a "luxury brand" with a "distinctive British" appeal, which makes a profit of £366m on a turnover of £1.86bn a year, all the while shedding British jobs and closing British factories has a stomach churning quality of its own' (Cadwalladr 2012: 1). Burberry countered that it:

> is clearly not pulling out of Britain, Burberry is not trying to in any way change its heritage … Burberry is one of the few remaining British retailers and manufacturers that still manufacture in Britain, and it is a huge success story for this country. It is very sad indeed that we just could not make the factory in Treorchy viable … [Burberry] is not just a British brand from a marketing sense; the heart of Burberry is here in Britain. In London we have all the design team, the marketing teams; it is the epicentre of what Burberry does globally. (John Peace, Chairman, Burberry, quoted in Welsh Affairs Select Committee 2007: 25, 3)

Figure 5.4 'Keep Burberry British' campaign demonstration, Burberry Store, New Bond Street, London. Source: Keep Burberry British Campaign http://keepburberrybritish.typepad.com/photos/campaign_photos/18112006010.html, accessed 20 November 2014.

With fashion brands intensifying their 'focus on the core elements of marketing and brand building while leaving increasingly larger parts of the value chain to other specialist countries' (Kwong 2008: 6) while balancing growing concerns about the carbon footprints of international trade, Burberry's longer term sourcing strategy from Britain remained in question. Although producing the iconic trench coat for the less cost sensitive and premium priced segment of the brand pyramid, Burberry's Rotherham factory in South Yorkshire was closed in 2009 with the loss of 174 jobs and its production transferred to Castleford following the collapse in demand after the 2008 global financial crisis and economic downturn.

Tracing the material geographical associations in the national origination of Britishness in the Burberry brand utilized by the participant actors demonstrates the integral importance and role of the brand and its branding in creating and shaping its uneven spatial circuits of meaning and value. This 'manufacture of meaning' (Jackson *et al.* 2007) is not only cultural construction by the actors involved; it is the creation of economic value by a brand owner and managers. This cultural and economic connection is demonstrated in the role that the tighter control and reinvigoration of

Burberry's geographical associations with Britishness played in its commercial turnaround from the late 1990s. A selective and particular version of the origination of Britishness in the brand and its branding was utilized in efforts to (re)create value and meaning that generated tensions in the uneven geographical development of the brand's production. Burberry's centralization in London reinforced the 'hierarchical time-space relation' of the stylistic and temporal dominance of the 'Paris-Milan-London-New York fashion axis' (Weller 2007: 60). Pressures for enhanced cost savings and financial returns transmitted through Burberry's plc ownership structure encouraged the management's integration and rationalization of its supply base in line with the efficiency savings and scale economies of global sourcing models. The brand pyramid and its particular geographical associations deepened our understanding, qualifying any simplified and economically determinist explanation and challenging singular accounts of solely brand-led 'retailers *without* factories' (cf. Klein 2000, emphasis added). But, as its material production geographies shift, focusing upon the brand reveals that 'what is made in Britain is the creation of style, the putting together of style, the advertising campaigns, the combination of detailing, the design service sector side of things is still made in Britain' (Professor of Design, author's interview, 2008). As a nationally framed and originated 'British' brand tensions remained for Burberry in accommodating the spatial dissonance between the international outsourcing of its material production and the enduring need for branded design and styling to provide the differentiation and distinctive meanings and value constituted by its particular geographical associations.

Circulating Britishness in Burberry

Originated by its owner as a 'British global luxury brand' (Burberry 2008: 50), the geographical associations of Britishness are central to the actors' attempts to construct, cohere and circulate value and meaning in branding Burberry in particular spatial and temporal market settings. This is because 'British brands ... have a design philosophy that transcends cultural differences and ... British design and British made is still something that resonates globally and comes with a price value that people are willing to pay' (Fashion Industry Export Association Official, author's interview, 2008). Despite underpinning its steady post-war growth, Burberry's particular version of the nationally framed origination of Britishness became marked as 'dowdy' (Watts 2002: 1) with limited innovation, 'churning out the same lines every year' (*The Economist* 2001: 1), and undermined by a 'moribund image' (Moore and Birtwhistle 2004: 414). Weakly controlled brand extension amongst wholesale and retail licensees in Europe and Asia during the 1980s had widened the product range and international distribution, but confused and diluted the brand's meaning and value through branding a wide array of merchandise, including coasters, umbrellas and whisky (Moore and Birtwhistle 2004). Geographically extended brand circulation in different market segments generated sales but 'was slowly killing the brand' (*The Economist* 2001: 1). By the 1990s, the Burberry brand had become 'weary and outmoded (Tungate 2005: 160).

Central to the late 1990s commercial turnaround was the dramatic modernization and tighter control of the brand's image and the particular variety of the origination of Britishness in its circulation:

> In keeping with efforts to enhance the brand's consistency across consumer touch points, all Burberry visual imagery is now derived from a single source – whether print advertisement, catalogue or in-store display ... synchronised to appear with the flow of related products to stores. (Burberry 2007: 45)

Reflecting the transition to a segmented brand pyramid, the strategies of Burberry's circulators sought to 'present a mix of products that present the overall brand image and which demonstrate the extent of the product range' (Moore and Birtwistle 2004: 420). Informed by its brand hierarchy, central artistic direction and uniformly consistent branding guided its reorganization:

> Burberry was a collection of decentralised regions and business units, each with an independent approach in areas such as design, merchandising and supply chain. Over the last year, continued progress has been made in working as one company and one brand – with a more cohesive global advertising campaign, small apparel licences brought in house and further strengthening of the regional and corporate teams. (Burberry 2008: 59)

The new, more proactive circulation strategy to 'integrate anything branded Burberry around the world' (Burberry 2009: 31) was manifest in a range of new activities in which articulating the origination of Britishness of the brand was integral. Reflecting the need for an international reach, the brand was renamed from 'Burberry's (of London)' to 'Burberry', removing the literal geographical association in its brand name. A step that branding commentators recognize as important: 'Removing the apostrophe and the "s" implies a change in the lettering as whole, modernizing the brand, doubtless internationalizing it, but perhaps slightly destabilizing part of its Anglo-Saxon clientele' (Chevalier and Mazzalovo 2004: 36). Further new circulation practices comprised: the deployment of reputable fashion photographers (e.g. Mario Testino); the utilization of celebrity models personifying the aspirant age, ethnicity and gender of the brand (e.g. British model Kate Moss); fashion shows focused upon generating media images shaped by brand hierarchy and place (e.g. Prorsum in Milan, Burberry London in London); brand web site enhancement (e.g. live streaming of fashion shows); and cost-effective high profile brand association and placement amongst global celebrities (e.g. Madonna, David and Victoria Beckham) (Tungate 2005). Such techniques for generating editorial – or so-called 'advertorial' – coverage amongst international media channels echoed the socio-spatial history of Burberry's pioneering geographical associations, for example in equipping Amundsen's 1911 South Pole expedition and exposure in cinema classics such as Audrey Hepburn's clothing in the New York depicted in *Breakfast at Tiffany's*.

The new circulation strategy tapped into and perpetuated the 1990s international media and public obsession with celebrity (Frank 1998). Burberry's twice yearly advertising campaigns in signature black and white sought to construct and cohere meaning and value through a particular British origination. These efforts comprised the iconography of the London black taxi, red telephone box and the model sons and daughters of British 'icons' (e.g. Bryan Ferry, Jeremy Irons): 'the look we wanted to achieve was a mix of Beaton imagery and the '60s feeling in London, reinterpreted with the people and the vigour of today. It's the new swinging London' (Mario Testino quoted in *Brand Republic* 2007: 1). This forging of revitalized geographical associations to originate the Britishness of the brand appeared to work commercially for the owners and managers as Burberry's acknowledged revival and contemporary repackaging 'brought an unexpectedly rebellious, streetwise image to the brand' (Tungate 2005: 161), hinged upon a renewed 'sense of 'fashionability' (John Peace, Chairman, Burberry quoted in Welsh Affairs Select Committee 2007: 3). Tight control of circulation was extended beyond Burberry branded goods and services to corporate external affairs for the Burberry Group. A narrative was articulated of successful contribution to the creative economy and British fashion (e.g. participating in national government Fashion Summits and export drives), and membership of British business lobbying groups (e.g. The Walpole, UK Fashion Exports).

Growing international circulation generated new tensions and accommodations in the geographical associations being utilized by the brand managers and circulators in efforts to originate the Britishness at the heart of the brand's revived meaning and value. First, spatial discontinuities were apparent between the geographically associated meanings and perceptions of its branded products and their material production geographies. The nationally framed, albeit highly selective and pliable, origination of Britishness constitutes the integral meaning and value of the Burberry brand prominent in its circulation. But as an internationalizing group deploying global sourcing models in the face of competition, not all of the brand's products are actually 'Made in Britain'. As explained above, the procurement strategy and decisions of the brand owner are conditioned by multiple concerns including authenticity, cost, quality, materials and brand hierarchy. In circulation, Burberry follows labelling regulations concerning country of origin of manufacture in specific market times and spaces and, depending upon its value and position, represents the brand's provenance in specific ways. Indeed, while the 'Keep Burberry British' campaign argued that "If Burberry want to be seen as a British brand they should not be sending these jobs overseas" (Charlotte Church, quoted in *BBC News* 2007: 1), Burberry claimed that its customer base is highly internationalized and little expectation existed that all its branded products would be exclusively 'Made in Britain' (John Peace, Chairman, Burberry quoted in Welsh Affairs Select Committee 2007). The particular socio-spatial history of the brand exerts a powerful and enduring influence upon the perceived meaning and value of the origination of Britishness almost irrespective of its material geographical associations of production. Fashion industry analysts, for instance, argued that 'I don't think Burberry has to be made in Britain in order for it to be British … the heart of the brand is British' (author's interview, 2008).

Second, the Burberry management faces a tension between the economic rationales of wider international distribution, accessibility and sales against the cultural-economic imperatives of restricting access to maintain the brand's exclusivity, cachet and premium price. While 'rooted in its authentic British heritage and the integrity of its outerwear', the actors seek to 'elevate and extend the brand' through its 'broad consumer appeal across genders and generations: a unique demographic positioning within the luxury arena and global reach' (Burberry 2008: 57). International distribution aims to fuel demand in existing core but slow growing market contexts (e.g. Japan, Spain), often amongst new aspirant social groups, 'under-penetrated' markets (e.g. United States, France and Italy) and emergent and growing markets (e.g. China, India, the Middle East and Russia) (Burberry 2008: 59). Indeed, this strategy helped the Burberry Group to weather the worst effects of the global financial crisis from its onset in 2008. However, widening circulation to 'downscale' socio-spatial groups risks diluting the brand's meaning and value for its 'upscale' customers and potentially displacing their purchases of higher margin branded products because 'the people who have got the money just do not want to be associated with those kinds of people' (Fashion Industry Journalist, author's interview, 2008). Burberry management attempted to handle this dilemma by constructing hierarchical market segments with clear originations in their geographical associations with Britishness. Prior to and following the 2008 global financial crisis and recession, these spatial attachments were being disrupted by the brand's need as a plc to capitalize upon the sheer size and rapid growth of the luxury goods market, estimated to be worth €170 billion and growing at 10% annually (Kwong 2008).

The geographical associations of the origination of Britishness play an integral role in shaping and, in turn, being shaped by the agency of circulators of value and meaning in the brand and branding of Burberry. Modernization and commercial revitalization focused upon more tightly controlled circulation of a particular, refreshed version of the origination of Britishness central to the Burberry brand, represented and articulated in material, symbolic and discursive forms. In this way, the brand has 'not just been repositioned, but "re-imagined"' (Tungate 2005: 161). The actors involved have reworked its 'Britishness ... defined sartorially through connotational codes to do with high quality, fine finishing, exacting standards and high monetary value' with 'class aspiration ... promoted as a key value-adding characteristic in the selling of British fashion and in British fashion exports' (Goodrum 2005: 129). Focus upon the origination of the Burberry brand and its branding elucidated the tensions and accommodations involved in the geographical associations integral to the geographically uneven development of its spatial circuits. Spatial discontinuities were evident in the relations between the circulated geographical associations and cultivated perceptions of origination and provenance of 'Made in Britain' with the internationalizing economic rationales of material production. The growth of wider international distribution raised issues for maintaining the brand's cachet of luxury exclusivity and premium pricing.

Consuming Britishness in Burberry

From its pioneering construction of a market for outerwear amongst local sportsmen in Hampshire in 1856 to its first shop in London's West End in 1891 (Tungate 2005), the origination of the Britishness of the Burberry brand and its branding has underpinned its meaning and value in consumption. The 'aspirations, connotations and … emotional attachment to certain places' (Fashion Industry Commentator, author's interview, 2008) establishes its meaning and value in the construction of economic and social identities: 'When buying a Burberry product, customers associate themselves with the heritage and craftsmanship of the brand' (Burberry 2008: 69). The origination of Britishness provides Burberry's managers and marketers with a meaningful and valuable history that can be 'drawn on, created … respect, longevity, history … it grounds a brand … it places it … gives it that depth, patina and set of associations from which marketers can select' (Fashion Industry Analyst, author's interview, 2009). Burberry's historic associations with specific products have provided core assets for the modernized (re)interpretation of its heritage. Use of such resources has conditioned its transition from 'fading raincoat manufacturer into a leading luxury goods brand' (Dickson 2005: 1) and enabled the extension of its product range and market reach. The 'brand icon' trenchcoat, for example, was adapted from the military officer's coat for a War Office commission in 1914 and became 'this iconic garment … even more popular after the war, sported by explorers, plain-clothes policemen, and members of the public with secret dreams of heroism' (Tungate 2005: 160).

Through Burberry's brand-led modernization, the managers and marketers have sought to uphold the consumption status and value while updating the meanings of its signature garments as design classics and continuity products with long life spans, sold year after year. The aim has been to ensure income streams are balanced by careful extensions and responsive innovations reflective of fashion trends (Moore and Birtwistle 2004). The particular renewal of the origination of Britishness has provided a differentiated and distinctive positioning in terms of price and style in competition against other fashion brands (Figure 5.5). Indeed, Burberry's successful turnaround and weathering of the economic downturn from 2008, recognized internationally as the top exporter of Britishness to the world (Woods 2006), established a business template that other historic British brands such as Aquascutum, DAKS and Dunhill have since struggled to follow.

The brand pyramid shapes the strategies of Burberry's managers and marketers for generating sales and purchases to enable consumption of the brand. At its apex, the Prorsum collection has limited and geographically specific distribution and premium prices as the 'couture/high fashion range' and:

> focus for fashion shows and editorial interest/coverage. Produced in limited quantities in order to satisfy the demand for exclusivity among affluent consumers … distributed through Burberry's flagship stores as well as through prestigious department stores including Barney's in New York and Harvey Nichols and Harrods in London. (Moore and Birtwistle 2004: 415)

The tailored Burberry London (International and Localised) collections have relatively wider distribution channels, geographies and lower price points. Historically, the casual Burberry Lifestyle diffusion collection has dominated sales by volume. The new strategy is aimed at 'enhancing the luxury quotient of the business' and 'leveraging the franchise' (Burberry 2008: 49, 59) to maximize the meaning and value of the

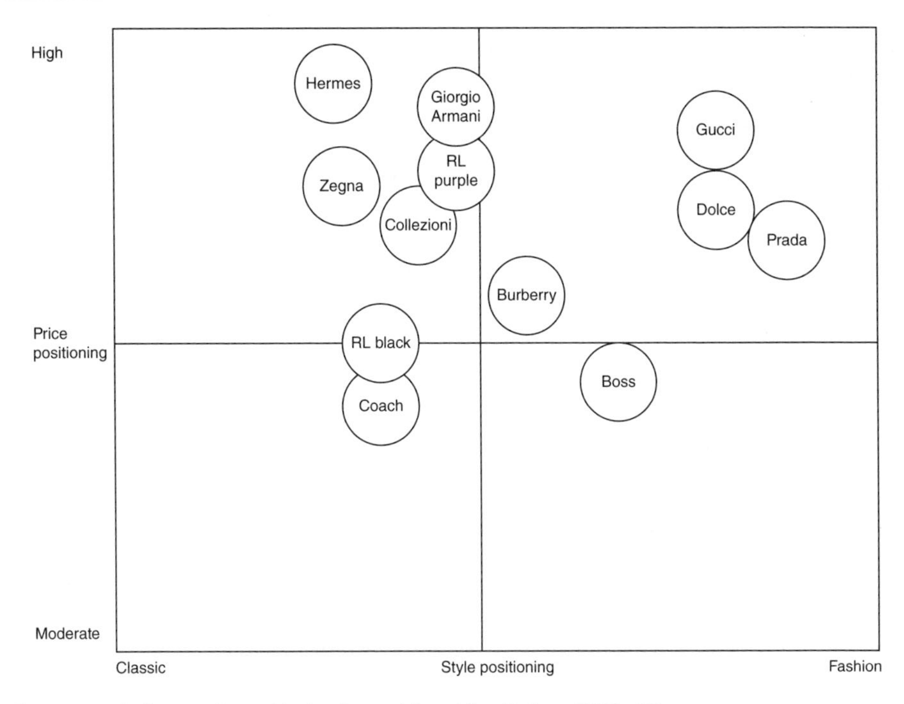

Figure 5.5 Burberry market positioning. Source: Adapted from Burberry (2006a: 31).

brand through its branding and reshape the brand pyramid towards the higher price and gross margin garments from the Prorsum and London collections.

The meaning and value of the origination of Britishness has propelled the internationalization of the consumption of Burberry, demonstrating 'the brand's appeal and cachet across cultures' (Burberry 2008: 11). Burberry's initial internationalization unfolded during the British Imperial era in exports to Europe, North and Latin America in the 1890s sold through retail stockists in New York, Buenos Aires and Montevideo, and its first overseas store in Paris in 1910 (Moore and Birtwistle 2004; Tungate 2005). Sustained but steady growth of the brand assumed a particular character, masking several problems until the 1990s. First, Burberry became heavily reliant upon a small array of core products and narrow consumer base of 'middle-aged, fashion conservative men' (Moore and Birtwistle 2004: 414). Second, loose and insufficiently discriminating distribution and licensing meant the brand was available in large numbers of small, inappropriate and low quality retail outlets internationally. This weak control allowed parallel, 'grey market' trading of Burberry branded products by legitimate wholesalers to other non-approved distributors and stockists, sold at discounts in key Asian markets and even re-imported for discounted sale in western markets. Last, dependence upon sales in the growing east Asian and especially Japanese markets meant the late 1990s Asian financial crisis impacted heavily, reducing profits from £62 million to £25 million in 1997–1998 and triggering GUS's

Table 5.1 Burberry Group, total revenue by area (£m), 2001–2009[a]

Year	*Total revenue*	*Europe*	*%*	*North America*[b]	*%*	*Asia Pacific*	*%*	*Rest of the world*	*%*
2001	427.8	259	60.5	90.9	21.2	74.6	17.4	3.3	0.8
2002	499.2	286.7	57.4	110.5	22.1	100.1	20.1	1.9	0.4
2003	593.6	302.7	51.0	140.5	23.7	147	24.8	3.4	0.6
2004	675.8	346.8	51.3	162.4	24.0	162.6	24.1	4	0.6
2005	713.7	356.4	49.9	164.1	23.0	186.6	26.1	6.6	0.9
2006	661.8	325.6	49.2	177.9	26.9	144.6	21.8	13.7	2.1
2007	764.2	381.6	49.9	196.5	25.7	167.5	21.9	18.6	2.4
2008	910.6	453.4	49.8	234.8	25.8	189.1	20.8	33.3	3.7
2009	1118.9	524.3	46.9	304.7	27.2	240	21.4	49.9	4.5

Note: [a]Nominal prices; [b]Americas category from 2005.
Source: Calculated from Burberry (2005, 2009).

installation of new management to reinvigorate the brand prior to its flotation as a plc (Finch and May 1998).

The renewed and modernized origination of Britishness was integral to the brand management's attempts to regain control over the how the brand was consumed from the late 1990s: 'Our goal is not to be Hermés or Bottega Veneta ... Britishness is so much a part of what we are about – now let's do that better than anyone in the world' (Rose Marie Bravo, quoted in Gumbel 2007: 124). The management strategy tried to accelerate growth and generate higher margins through a 'fundamental shift in the ... operating culture' to become more 'consumer-centric' and 'responsive' by a 'move from a relatively, static, traditional wholesale structure to a more dynamic, retail culture and mindset' (Burberry 2008: 26). Reorganizing the distribution network sought 'to ensure a high quality consumer experience at every stage of the sales process ... embedded within the business's systems, policies and processes from our branding policy, fabric guide manual and wearer trial policy, through to the measurement of customer satisfaction' (Burberry 2005: 38).

The strategic shift involved, first, internationalization of the brand's sales and articulation of a broader 'global' reach. Europe doubled its total revenue to £524 million but reduced its share from 61 to 47% between 2001 and 2009 while North America/Americas rose from 21 to 27%, Asia Pacific increased from 17 to 21% and the rest of the world grew from 0.8 to 4.5% (Table 5.1). Tellingly for the brand given its strong origination in Britishness, 'for brands like Burberry, the home market almost becomes marginal, it is not where the money comes from' (Fashion Industry Analyst, author's interview, 2008). Second, the retail channel for the brand's distribution rose from 33 to 52%, while wholesale and licensing were reduced from 56 to 41% and 11 to 7% respectively between 2001 and 2009 (Table 5.2). This rebalancing aimed to ensure retail provided the 'impetus for media and consumer interest in the Burberry brand within respective markets which precipitate wholesale sales, while the profits from wholesaling ensure that flagship stores are economically viable' (Moore and Birtwistle 2004: 419). As part of this transition, the Burberry Group had to absorb the short

Table 5.2 Burberry Group, total revenue by channel (£m), 2001–2009[a]

Year	*Total revenue*	*Wholesale*	*%*	*Retail*	*%*	*License*	*%*
2001	428	239	56	143	33	46	11
2002	499	289	58	157	31	54	11
2003	594	307	52	228	38	58	10
2004	676	351	52	257	38	67	10
2005	716	372	52	265	37	78	11
2006	743	343	46	319	43	81	11
2007	850	354	42	410	48	86	10
2008	995	426	43	484	49	85	9
2009	1202	489	41	630	52	83	7

Note: [a]Nominal prices.
Source: Calculated from Burberry (2005, 2009).

term costs of reduced sales revenue from not renewing licenses in key markets, amounting to £1.7 million in certain menswear contracts (Burberry 2008: 62). Last, 'intensifying non-apparel development' (Burberry 2008: 11) involved brand extension to widen and diversify the brand's product range against its more broadly based competitors, especially in higher margin accessories, furnish its flagship stores and populate its lifestyle branding offer (Moore and Birtwistle 2004). Non-apparel quadrupled in revenue to £366 million and grew as a proportion of revenue from 23 to 31% between 2001 and 2009 (Table 5.3).

Reflecting the claimed shift towards the consumption of meaning and value through experiences rather than just the possession of objects (Pine and Gilmore 1999), Burberry's new retail culture is reflected in the emphasis upon so-called 'consumer touch points' – runway, advertising, editorial, PR, windows, in-store, on-line – for the brand. Replacing the replication of retail concepts across distribution channels, this geographically differentiated strategy emphasizes:

> change in store design and even product offering by location as we look to pre-empt customers' ennui from seeing the same thing in London, New York and Tokyo. ... Customers will want to have different, fresh experiences when they shop. Companies must therefore work on shorter cycles: putting together more 'capsule collections' of clothing and accessories and delivering them more often to create the suggestion to customers that if they don't buy it now, it may not be there later. (Rose Marie Bravo, former Chief Executive, Burberry, quoted in *The Economist* 2005: 1)

Accelerating retail-led growth has been delivered through the geographical expansion of directly operated stores, from 292 to 419 between 2007 and 2009, especially in under-penetrated markets, including the Americas, Middle East, China, India and Russia (Figure 5.6). The retail outlets aim to support the brand message and its core, aspirational values. Flagship stores have been placed alongside other brands in particular shopping districts in key markets – such as Bond Street,

Table 5.3 Burberry Group, total revenue by product (£m), 2001–2009[a]

Year	Total revenue	Womenswear	%	Menswear	%	Non-apparel[b]	%	Other[c]	%	Licensing	%
2001	427.8	134.7	31.5	142.4	33.3	98	22.9	6.9	1.6	45.8	10.7
2002	499.2	165.2	33.1	149.4	29.9	125.8	25.2	5.3	1.1	53.5	10.7
2003	593.6	197.9	33.3	162.8	27.4	169.5	28.6	5.1	0.9	58.3	9.82
2004	675.8	225.7	33.4	190.1	28.1	189	28.0	4	0.6	67	9.91
2005	728.1	242.1	33.3	194.5	26.7	197.6	27.1	15.5	2.1	78.4	10.8
2006	742.9	249.3	33.6	206.2	27.8	189.2	25.5	17.1	2.3	81.1	10.9
2007	850.3	305.5	35.9	227	26.7	211.2	24.8	20.5	2.4	86.1	10.1
2008	995.4	345.2	34.7	247.8	24.9	289.7	29.1	27.9	2.8	84.8	8.52
2009	1201.5	412.8	34.4	298.4	24.8	366.3	30.5	41.4	3.4	82.6	6.87

Note: [a]Nominal prices; [b]including childrenswear until 2004; [c]including childrenswear from 2005.
Source: Calculated from Burberry (2005, 2009).

London; Fifth Avenue, New York; and Rue Saint-Honoré, Paris – often as costly loss leaders cross-subsidized for their contribution to the broader group through projection of brand image, power and status (Power and Hauge 2008). Elsewhere, concessions reflect national consumption preferences and have been concentrated amongst prestigious retailers in Japan, South Korea and Spain. The origination of Britishness suffuses retail outlets for the Burberry brand, manifest in their décor, styling, in-store visuals and in the recent introduction of bespoke, made-to-order services in seeking to 'enhance the consumer's experience of the Burberry brand' (Watts 2002: 1). Yet such cultural construction of meaning and value in the consumption of the brand has material bases and strong underpinning economic rationales focused upon 'Enhancing store productivity. … Through balanced assortments, compelling visual presentations, improved sales and service and engaging marketing … to drive traffic, increase conversion and build average transaction size' (Burberry 2008: 32).

The geographical tensions for actors in accommodating availability versus exclusivity in producing and circulating the origination of Britishness in the Burberry brand are replayed in its consumption. For consumers of fashion, originating the brand in a nationally framed Britishness creates meaning and value as a:

> rationale behind a product that claims to be deluxe. In a mature economy, a consumer's self-confidence derives from being discerning rather than merely rich. Subtle details … add depth to the product experience. … This leads luxury brands to revert to craftsmanship, detail, creativity, and innovative, show-like advertising. (Ricca 2008: 67)

It is less clear whether and to what extent the Britishness of Burberry branded goods and services is valuable and meaningful for consumption because of their perceived *or* actual symbolic geographical associations and material origins. For some fashion

Figure 5.6 Burberry store locations, 2006[a]. Source: Adapted from Burberry (2006a: 41).
Note: [a]Includes franchise stores.

industry journalists, 'you can't help feel less affection towards a British brand that isn't produced in Britain. I would feel less keen to buy ... if ... I realized it was made in China not Yorkshire. ... It wouldn't feel so authentic' (author's interview, 2008). In contrast, although confusing the geographical associations of 'UK' and 'England', McLaren *et al.* (2002: 41) argue that 'a successful UK brand, such as Burberry or Aquascutum, would not be excessively damaged through being physically made outside the UK' and cast 'doubt on the value of a "made in England" label in the fashion sector as a whole'. Some fashion business analysts go further in this view claiming that, in the wake of socially unequal growth and emergent markets, 'we have a global rich ... that ... might not care, or have any clear understanding of the origination of those products' (Professor of Marketing, author's interview, 2008). Others note the importance of the 'perception of where the brand comes from, the national identity of the brand' because 'even though the merchandise is not physically made ... in those countries the fact that it is a British brand and promoted as such is perceived added value and does add dramatically to the price you can charge for a product. Even if it is made in China' (Fashion Industry Export Association Official, author's interview, 2008). Indeed, rather than a 'story of brand inconsistency and untold damage to a company's reputation', branding commentators argued that the Treorchy closure and 'Keep Burberry British' campaign has had negligible if not positive impact on Burberry's sales in balancing 'the ocean of heritage with a constant stream of contemporary notoriety' (Ritson 2007: 1).

The 'democratization of luxury' (Silverstein and Fiske 2003) driven by consumers trading up from mainstream to 'luxury' brands embodying greater emotional value has further propelled the international distribution and growth of sales for the Burberry brand. This growth has sharpened the concern about managing the exclusivity and prestige of the controlled brand access and distribution used by participant actors to construct the scarcity underpinning its premium pricing. While tightening control reinvigorated the brand following its relative decline during the 1980s, Burberry's luxury cachet is under pressure from its current growth and expansion as it seeks to become 'a "truly global" luxury brand' (Angela Ahrendts, Chief Executive, Burberry, quoted in Callan 2006: 1). Widening distribution risks Burberry becoming only a 'premium brand ... able to distinguish themselves and charge a premium price' (Fashion Industry Analyst, author's interview, 2008) relative to more exclusive luxury brands such as Chanel and Louis Vuitton able to command much higher price premiums and profit margins.

Current plans crystallize the brand owner's dilemma about 'moving downmarket to find growth' and risking exposure to cyclical market segments (Milne 2008: 2), such as rapid American retail outlet expansion in lower tier urban centres such as Kansas, Indiana and Ohio. Similarly, increased levels of brand extension into high margin, branded accessories may enrol new generations of consumers but question whether the 'clothes simply become the promotional tools for branded goods' (Tungate 2005: 146). Burberry managers have tried to utilize the brand pyramid for market segmentation. But for some industry analysts 'they have just got all these brand identities that don't play clear, distinct and separate roles under one umbrella' (Fashion Industry Analyst, author's interview, 2008). It is uncertain too how enduring and stable the hierarchies the Burberry managers have developed to structure the brand and its sub-brands will turn out to be. Disruptive forces threatening to undermine the coherence of particular articulations of the brand are evident. First, boundaries are blurring in specific spatial and temporal market contexts and new and hybrid categories and segments are emerging, for example, in streetwear, sportswear and 'semi-couture' (Tungate 2005). Second, consumer agency is active in mixing and combining items from different mass and prestige brands and segments in search of 'individual style – where everything is personalised and customised' (Rose Marie Bravo, former Chief Executive, Burberry, quoted in *The Economist* 2001: 1).

Uneven spatial circuits of value and meaning in the consumption of the brand and branding of Burberry are geographically associated with the particular, modernized version of the origination of Britishness constructed and articulated by its owners, managers, circulators, consumers and regulators. Shaped by the brand owner's hierarchical market segmentation of its brand pyramid, attempts at a careful renewal of historically resonant meaning have underpinned continued valuation of Burberry *vis-a-vis* its competitors amongst luxury fashion consumers. Exactly how such 'luxury' and high margin such assessments continue to be has become a concern. Predicated on its distinctive and nationally framed origination of Britishness, brand-led sales growth has underpinned Burberry's reconfigured and more tightly controlled retail focus and internationalization since the late 1990s: unevenly expanding its global spread of stores in emerging and growing markets; rebalancing its distribution channels from wholesale to retail; and developing higher margin brand extensions. Patterns of the brand's consumption reveal mixed and inconclusive evidence of the impact of geographical discontinuities in its perceived and actual geographical associations and material origins. Given its international sales growth, purchasers in Dubai, Durban and Dalian appear to be consuming the Britishness of Burberry with apparently little regard for where the branded goods are actually made or branded services delivered from. The actors' origination of Britishness and its particular meaning and value seem to have constructed and fixed sufficient perceived authenticity to guard against the risks that sophisticated luxury fashion consumers will 'lose trust for brands that have only surface-level offerings' (Okonkwo 2007: 92) and be less willing to pay premium prices for Britishness in brands materially produced beyond the national territory of Britain. This tension in cohering the nationally framed origination of the brand constitutes an ongoing challenge for the actors involved in trying to balance accessibility and exclusivity in the brand's consumption in a phase marked by growth, internationalization and extension, buffeted by coping with and responding to disruptive and geographically differentiated shifts in particular spatial and temporal market contexts.

Regulating Britishness in Burberry

Since 'the Burberry brand ... is arguably one of our most important assets' (John Peace, Chairman, Burberry quoted in Welsh Affairs Select Committee 2007: 26), it is strongly regulated and protected by its owner and managers. The nationally framed origination of Britishness constituting the brand's source of meaning and value is captured in the Intellectual Property Rights and 41 trademarks owned by the Burberry Group. Ownership and valuation of these devices are key financial assets that shape the investment community's judgements and sentiment in making decisions about future growth prospects, share purchases and sales, and access to capital. Indeed, the strength of Burberry's reputation and value during its post-war growth underpinned its acquisition by retail and catalogue conglomerate GUS in 1955, which funded its initial retail network expansion in the United Kingdom and America (Moore and Birtwistle 2004). By the 1980s, the accumulation of problems documented above undermined the coherence of the particular origination of Burberry's meaning and value so that by the late 1990s financial analysts rated the brand as 'an outdated business with a fashion cachet of almost zero' (quoted in Finch and May 1998: 1). Burberry's profits collapse triggered new management and brand-led reinvigoration.

Reflecting a wider financialization of fashion and the ways in which luxury fashion brands are 'the engine of the entire business model ... responsible for most of the value created by their companies' (Ricca 2008: 66), Burberry was demerged from GUS to create a separate company once the commercial revitalization began showed signs of success in 2002. The new entity was floated on the share market in London as a publicly limited company 50 years after its acquisition. The aim was to establish a market value for Burberry for GUS shareholders and to build upon its business and financial momentum. Flotation integrated the brand owner more tightly into the capital markets in the City of London and, unlike its competitor groups that have retained large and influential blocks of family ownership, the orbit of its major institutional and international shareholders – including Blackrock

Inc. (9.94%), Schroders Plc (8.49%) and Massachusetts Financial Services Company (5.94%) (Burberry 2008: 57). Institutional ownership and intensified media and financial community scrutiny focused the Burberry brand managers upon raising sufficient capital to fund its international expansion and embedded the imperatives of increasing growth, share price and dividends and the fiduciary duties of shareholders to maximize investment returns firmly into its strategy because 'like in other publicly managed enterprises, when you are managing for quarterly results, it constrains the way you manage' (Fashion Industry Analyst, author's interview, 2008).

The brand and its nationally framed origination of Britishness is a central asset for the Burberry Group actors in their efforts to create and fix value and meaning to generate increases in key financial indicators, reflected in the strategic priority of 'leveraging the franchise'. The Group's brand-led resurgence has propelled growth in revenues to over £1 billion. Leading up to the 2008 global financial crisis and economic downturn, profits increased to over £200 million, market capitalization reached £2.5 billion and the share price rose consistently, outperforming the market (Figures 5.7 and 5.8). The crisis events from 2008 were manifest in a dramatic reduction in profits, while sales turnover held up by demand in emerging markets. This financial success culminated in the Group's entry into the *Financial Times* and the London Stock Exchange (FTSE 100) listing of leading shares. Burberry's share price declined further than the market in 2007–2008 but recovered robustly from 2009 again fuelled by growth in emergent markets (Figure 5.8). Burberry's performance raised investor expectations concerning future performance. Such imperatives mediated through financial actors have shaped the quantitative and qualitative character of Burberry's growth, for example, intensifying its brand extension into non-apparel categories because 'These high-margin goods tend to withstand swings in consumer spending and are key to the premium investment ratios enjoyed by luxury groups such as the French LVMH and Hermès' (Callan 2006: 1).

Reflecting the origination of the brand's particular geographical associations in the meaning and value of Britishness and the Creative Director's employment history, the renewed brand-led business model being pursued by the Burberry Group is a hybrid version of American and European luxury fashion brands such as Ralph Lauren, Gucci and Prada. Such models adapt common business elements (e.g. senior and experienced American retail talent, 'star' Creative Directors, global sourcing) (Tokatli 2012a) with their specific institutional histories and characteristics, and the particular originations of their brands (e.g. America for Ralph Lauren, Italy for Gucci and Prada). For Burberry with its origination of Britishness integral to the meaning and value of its geographical associations, the 'brand new world' is marked by having to 'embrace the spirit of risk-taking and experimentation with the financial responsibilities of a commercial enterprise' and the 'balancing act' of offering:

> the customer a sense of discovery, a one-of-a-kind experience and true product innovation against this approach's inherent demands on human and capital resources, lower economies of scale and the resulting implications on profit. At the same time, this constant quest for the unique and the new must be pursued while keeping a sharp eye on

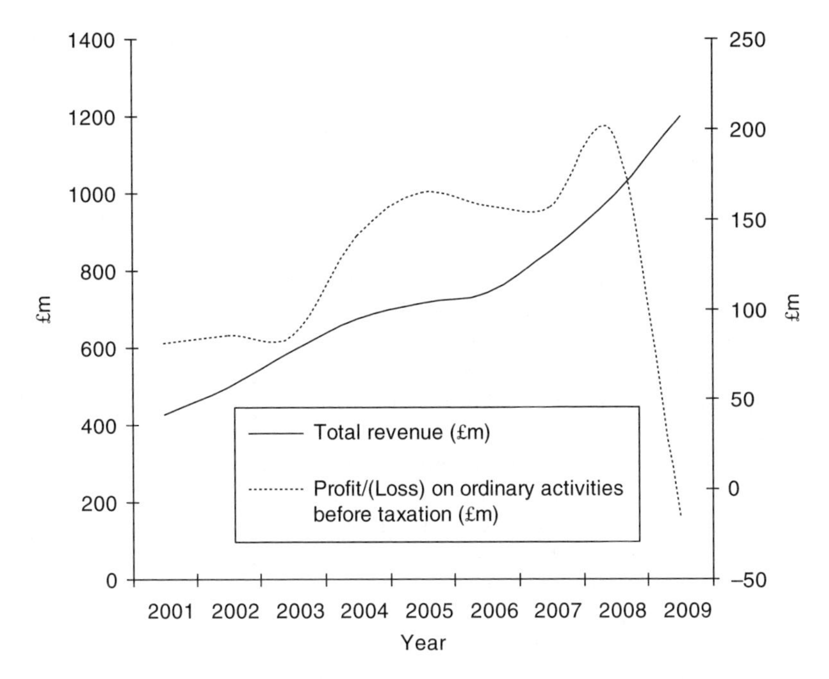

Figure 5.7 Burberry Group PLC, total revenue and operating profits, 2001–2009[a]. Source: Calculated from Burberry (2005, 2009).
Note: [a]Nominal prices.

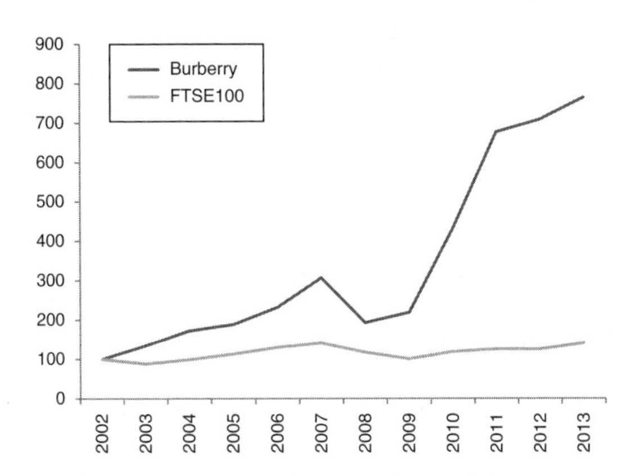

Figure 5.8 Burberry Group PLC share price and FTSE100 index, 2002–2009 (2002=100)[a]. Source: Calculated from Burberry share price and FTSE100 data; http://finance.yahoo.com/q/hp?s=BRBY.L&a=06&b=12&c=2002&d=06&e=31&f=2013&g=m;http://uk.finance.yahoo.com/q/hp?s=%5EFTSE&b=2&a=03&c=1984&e=31&d=06&f=2013&g=m.
Note: [a]Annual average from monthly adjusted closing price. Nominal prices.

> the heritage, history, integrity and credibility of a brand in an ever-changing and often unexpected competitive landscape. (Rose Marie Bravo, former Chief Executive, quoted in *The Economist* 2005: 1)

For some commentators, preserving authenticity is challenging because 'as soon as they all went public the drive for growth and return immediately scuppered any chance they had for retaining luxury status. The two of them are uncomfortable bed fellows' (Fashion Industry Analyst, author's interview, 2008). The business model being pursued by the Burberry brand's managers faces significant and intensified competitive risks too against 'international luxury goods groups who control a number of luxury brands and may have greater financial resources and bargaining power with suppliers, wholesale accounts and landlords. If Burberry is unable to compete successfully, operating results and growth may be adversely impacted' (Burberry 2008: 66). Indeed, despite the poison pill defence of a £200 million revolving credit facility and strategic concerns that 'no British brand ... has flourished under foreign ownership' (Fashion Industry Commentator, author's interview, 2008), the single brand-based growth and relatively small size of the Burberry Group led some to suggest it may become a takeover target for one of the larger 'multi-brand powerhouses such as LVMH and Gucci' (*The Economist* 2001: 1).

Reflecting how 'Design-based firms constantly struggle to prevent their aesthetic knowledge from leaking out to places where its economic value can be appropriated by imitators' (Weller 2007: 51), contestation over the regulation of the origination of Britishness in the Burberry brand and its branding are evident in the management's ongoing attempts to contain Burberry counterfeits. Fakes are seen as infringing trademarks and risks confusing and diluting the owners' particular attempts to create and fix the meaning and value:

> Burberry's trademarks and other proprietary rights are fundamentally important to the success and competitive position of the business. Unauthorised use of the 'Burberry' name, the Burberry Check and the Prorsum horse logo as well as the distribution of counterfeit products damage the Burberry brand image and profits. (Burberry 2009: 46)

The ambiguity of having a core brand value and icon 'recognised everywhere from Brazil to Japan' as an 'instantly recognisable brand signifier' (Haig 2004: 143) adversely affected the Burberry Check in the United Kingdom during the 1990s. The subcultural appropriation of Burberry by 'tough guys, skanks, soccer hooligans, lower-class unsophisticates, and cheesy celebrities' (Gross 2006: 1) reinterpreted the brand and associated it with a particular version and origination of Britishness as 'the ultimate symbol of nouveau riche naff' (Tungate 2005: 29; see also Power and Hauge 2008). This unwanted set of geographical associations by specific actors constructed an alternative origination of Britishness for the Burberry brand, stimulating the growth in counterfeits in ways that were perceived by its owners to be potentially damaging to the meaning and value of its central brand asset (Moor 2007). The impact of this episode reverberated all the way back to the Burberry Group's core design and branding, resulting in reduced use of the brand's key but easily replicable signature check, prominent logos and withdrawal of lower value brand entry level items (e.g. baseball caps).

The Royal Warranty of Burberry is a further concrete manifestation of the regulation of the particular nationally framed origination of the brand's Britishness. Awarded by Royal Appointment for valued suppliers of goods and services to members of the monarchy, Burberry holds Royal Warrants from Queen Elizabeth II (1955) and the Prince of Wales (1989). As a 'peerage for trade', the Royal Warrant is meaningful and valuable as a 'statement of quality, excellence, service, reliability' (National Secretary, author's interview, 2008). However, this form of regulation for the brand and the geographical associations of its branding is ambiguous and potentially disruptive to the brand's meaning and value in particular spatial and temporal market settings. The Royal Warrant geographically associates the brand with a particular version and meaning of the origination of Britishness in the monarchy, tradition and even empire valued by particular consumers, for example in Japan and the United States. But these same geographical associations hold less beneficial or even negative meanings for consumers that value the meaning of the origination of the Burberry brand in versions of Britishness that are characteristically 'cool', fashionable and modern. The Burberry brand managers attempt to accommodate these tensions by articulating the multi-faceted nature of its 'democratic brand' whereby 'parts of the brand would accommodate Kate Moss and others would appeal to an octogenarian monarch' (Burberry spokesperson, quoted in Godsell 2007: 1). Significantly, the relevant coats of arms and 'By Appointment' markers that can be used by holders of Royal Warrants are largely absent from the Burberry brand's circulation and promotion.

As a form of regulation of the geographical associations in a brand that connected it to a specific national territory, Royal Warranty was targeted by the 'Keep Burberry British' campaign, which questioned whether the brand's international outsourcing and closure of the Treorchy factory contravened the rules of the Royal Warrant by making the brand outside Britain:

> it is outsourcing its polo shirt production, at present in Treorchy, to other parts of the world, and unless it can guarantee that it will not be doing so on the basis of slave wages or child labour, it should lose its Royal Warrant. Specifically, if Burberry chooses to withdraw from Wales, it could be argued strongly that it should lose the Royal Warrant from the Prince of Wales. ... It is not for the Government to determine how Burberry should run its business, but it is for Parliament to consider how business is done by British companies that carry the national seal of approval – namely, the Royal Warrant. (Bryant 2007: 2)

The Burberry management countered that 'the Royal Warrants are important to Burberry; there is no condition that products sold by Burberry are manufactured in Britain but, clearly, Burberry still manufactures product in Britain, but from the point of view as a retailer, as a wholesaler, as a distributor of first class fashionable luxury products, then an endorsement via the Royal Warrants is very important to the company' (John Peace, Chairman, Burberry, quoted in Welsh Affairs Select Committee 2007: 19). In relation to the economic and social conditions under which its branded goods are produced as a result of international outsourcing, Burberry

Table 5.4 Burberry Group, employment by area, 2004–2009

Year	*Europe*	*%*	*North America*[a]	*%*	*Asia Pacific*	*%*	*Total*
2004	2657	68.7	747	19.3	465	12.0	3869
2005	2788	67.5	837	20.3	506	12.2	4131
2006	3066	65.9	902	19.4	683	14.7	4651
2007	3457	66.3	1026	19.7	735	14.1	5218
2008	3572	63.1	1339	23.7	749	13.2	5660
2009	3593	57.9	1616	26.0	999	16.1	6208

Note: [a]Americas category from 2008.
Source: Calculated from Burberry (2005, 2007, 2009).

further stated that 'Regardless of where the Group's factories are based they are governed by its ethical trading policy, which includes managing labour standards and environmental conditions in the factories that produce Burberry's luxury goods' (Burberry 2008: 69).

The material geographical associations to sites of production and employment for the Burberry brand demonstrate the tensions and accommodations between the meaningful and valuable national origination of Britishness in the territory of Britain and the relational extension of its geographies driven by its growing internationalization in search of future growth. As total employment in the Burberry Group has grown to over 6000, employment has been reduced in Europe from 69% to 58% of the total between 2004 and 2009 while North America/Americas and Asia Pacific have risen from 19 to 26% and from 12 to 16%, respectively (Table 5.4). In managing this tension between its originated Britishness and growing global operations and growth, the Burberry management stressed that it 'is an example of a truly global company. We are very much headquartered and based in Britain, almost half the workforce is based in Britain, all the design and marketing is here and yet less than 10% of our sales are in Britain, so we are the embodiment of globalisation' (John Peace, Chairman, Burberry quoted in Welsh Affairs Select Committee 2007: 7). The company further emphasized the central importance of its London Headquarters, stating that it:

> employs approximately 5,000 people worldwide of which 2,000 are located in Britain. In the past five years Burberry's British workforce has increased by 500 people (or over 30%), and half of this growth has been in manufacturing, especially of our high-value products such as the Burberry trench coat, which is produced at Castleford and Rotherham in Yorkshire. (Burberry 2007: 61)

Assessing the recent international growth of the Burberry brand, commentators argued 'It may not play well in Treorchy, but that success should still be labelled "Made in Britain"' (Hill 2007: 1).

The regulation of the origination of Britishness is central to the uneven geographical development of the spatial circuits of value and meaning of the Burberry brand

and its branding. As the key source of meaning and value creation and valorization as a corporate financial asset, the brand and its nationally framed geographical associations have shaped its changing ownership structure from part of the GUS conglomerate to a plc, integrating its evolution more closely with the capital markets and financial community in the City of London. This financialization of the brand has embedded the growth orientation of the Group's business model and challenged its management to cohere and safeguard its particular cultural meanings and value while delivering on economic imperatives in internationally expansive market times and spaces. The value and meaning of the brand in its origination in Britishness are protected through the regulatory devices of IPR and trademark as the:

> economic value of this proprietary fashion knowledge is created by states through their regulation of intellectual property rights, and is bounded by firms' brand identities ... this form of knowledge is purposefully 'de-placed' or universalised to maximise its penetration of geographical space while at the same time preserving the boundaries of its niche in aesthetic space. (Weller 2007: 44)

Subcultural appropriation and reworking in the United Kingdom triggered branding and design adaptation. Royal Warranty afforded further regulation of the Burberry management's particular origination of Britishness with ambiguous ramifications managed by emphasizing the multi-faceted appeal of the brand. The geographical associations in the brand regulated by Royal Warranty were questioned by the 'Keep Burberry British' campaign amid contestation concerning the uneven geographical development of the nationally framed origination of the Britishness of the brand in the wake of international outsourcing.

Summary and Conclusions

The nationally framed geographical associations of the origination of Britishness have been central to the efforts of producers, circulators, consumers and regulators in trying to construct and cohere meaning and value in the brand and branding of Burberry in the international fashion business. The particular socio-spatial history of the brand has provided a rich set of pliable assets and stories of geographical associations with Britain and Britishness. Yet the actors involved have had to engage the disruptive rationales of accumulation, competition, differentiation and innovation disturbing particular spatial and temporal market contexts in their attempts to construct and stabilize the origination of Britishness for the brand. Initially, the actors held together a particular version that underpinned and sustained its post-war growth but ultimately sowed the seeds of its commercial decline. The commercial turnaround and revival since the 1990s drew heavily upon the brand's origination in Britishness selectively to reinvigorate the brand and sustain its viability through the global financial crisis and economic downturn from 2008. Yet the contemporary growth of the brand has again encountered disruptive logics that threaten to undermine its stability as a competitive commodity able to deliver value flows and capital expansion in its spatial circuits. Segmentation and splintering have destabilized existing and created new market categories, and subcultural appropriation by particular consumer groups continue to intrude on the brand owner's specific origination of the meaning and value of the Burberry brand. Pressures for enhanced cost savings and financial returns transmitted through the Burberry Group's plc ownership structure have encouraged the integration and rationalization of its organization and operation. Origination advances our understanding by questioning overly simplified and economically determinist explanations, challenging Naomi Klein's (2000) characterizations of brand-led 'retailers without factories'. As a nationally framed, 'British' brand tensions remain for Burberry in accommodating the spatial dissonance between the international outsourcing of material production and the enduring need for branded design and styling to provide the differentiation and distinctive meanings and value of its particular origination of Britishness. Given its international growth in sales revenues, purchasers across the world appear to be consuming the Britishness of Burberry with seemingly little regard for where it is actually made. At least for now, actual and perceived authenticity appears to have been preserved against the risks that sophisticated luxury fashion consumers will be less willing to pay premium prices for Britishness in brands materially produced beyond the national territory of Britain. Origination explains how the material and relational geographies of the Burberry brand and its branding intersect to create a more complex and messy picture in which 'Consumers no longer expect clothes from British labels to be made in Britain, just as British designers are frequently defined by being trained and based here, rather than having British roots' (Long 2012: 10).

Chapter Six
'Global' Origination ... Apple

Introduction

Scrutinizing its portrayal as a somehow placeless and ubiquitous 'global' brand, origination reveals how the meaning and value of Apple in the shifting spatial and temporal market contexts of the international technology business have been inescapably geographically associated in Silicon Valley, California, by the actors involved. The brand and its co-founder Steve Jobs originate from the particular geographical context of Cupertino, Santa Clara County, California, in the heart of Silicon Valley – christened 'Computertino' in the late 1970s (Moritz 2009: 288). The attributes of the geographical associations of Silicon Valley as innovative, radical, revolutionary and youthful (Moritz 2009; Walker and the Bay Area Study Group 1990) have been used to imbue the brand with a sense of 'creativity, freedom of thought ... Silicon Valley values that people associate with the place ... sunny, optimistic, futuristic ... gold rush ... stick with rich associations ... Apple embodies all these things and especially Jobs' (Brand Web Site Editor and Publisher, author's interview, 2013). As a constructed and to a degree fluid spatial entity, 'Silicon Valley' can be configured as a local, sub-regional or regional scale and it performs an integral role as a meaningful and valuable place in the origination of the Apple brand. Apple's socio-spatial history demonstrates an origination at once relational and 'global' in its international reach *and* territorial and local, sub-regional or regional in its deeply rooted geographical associations in Silicon Valley, California. The Apple brand has 'cachet in coming from a particular place' (Brand Commentator, author's interview, 2013).

Origination: The Geographies of Brands and Branding, First Edition. Andy Pike.
© 2015 John Wiley & Sons, Ltd. Published 2015 by John Wiley & Sons, Ltd.

Producing the 'global' in Apple

The origination of the ostensibly 'global' brand of Apple can be traced back to its establishment in 1976 in what was coined 'Silicon Valley USA' by columnist Don Hoefler in trade publication *Electronic News* in 1971 (Isaacson 2011: 10) (Figure 6.1). College drop-outs Steve Jobs and Steve Wozniak set up the company with $1300 – about $4222 in 2012 money – in the Jobs' family garage in Cupertino, Santa Clara County, California. Steve Jobs was in and of Silicon Valley:

> You had all these military companies on the cutting edge. It was mysterious and high-tech and made living here very exciting ... [Frank] Terman [Dean of Engineering, Stanford University; see Saxenian 1996] came up with this great idea that did more than anything to cause the tech industry to grow up here. ... Growing up, I got inspired by the history of the place. That made me want to be part of it. (Steve Jobs, quoted on Isaacson 2011: 9–10)

Silicon Valley's strong geographical associations shaped Jobs' ethos and outlook, moulding the distinctive 'born-in-America gene for Apple's DNA' in the brand (Hartmut Esslinger, frogdesign, quoted in Isaacson 2011: 133). These 'foundation stories' for the Apple brand serve to 'connect the brand to the early days of Silicon Valley and the development of the personal computer' and 'tie the brand to the history and development of the industry, thereby connecting it to time and place' (Beverland 2009: 41).

These capabilities in Silicon Valley were produced by a decentralized and open industrial system and culture with diverse, multiple and constantly changing interconnections enabling and supporting new and existing entrepreneurs in emerging sectors (Saxenian 1996, 1999, 2005).

The hip, maverick and iconoclastic social and cultural mores of laid-back Californian directness geographically associated with a particular place and time in California marked the founding and branding of Apple:

> In San Francisco and the Santa Clara Valley during the late 1960s, various cultural currents flowed together. There was the technology revolution that began with the growth of military contractors and soon included electronics firms, microchip makers, video game designers, and computer companies. There was a hacker subculture – filled with wireheads, phreakers, cyberpunks, hobbyists, and just plain geeks – that included engineers who didn't conform to the HP [Hewlett Packard] mold and their kids who weren't attuned to the wavelengths of the subdivisions. There were quasi-academic groups doing studies on the effects of LSD. ... There was the hippie movement, born out of the Bay Area's beat generation, and the rebellious political activists, born out of the Free Speech Movement at Berkeley. Overlaid on it all were various self-fulfilment movements pursing paths to personal enlightenment: Zen and Hinduism, meditation and yoga, primal scream and sensory deprivation, Esalen and est. (Isaacson 2011: 56–7)

From its early beginnings, Steve Jobs and the founders of Apple tried to fix the brand with particular meaning and value 'forever associated with Cupertino, Silicon Valley

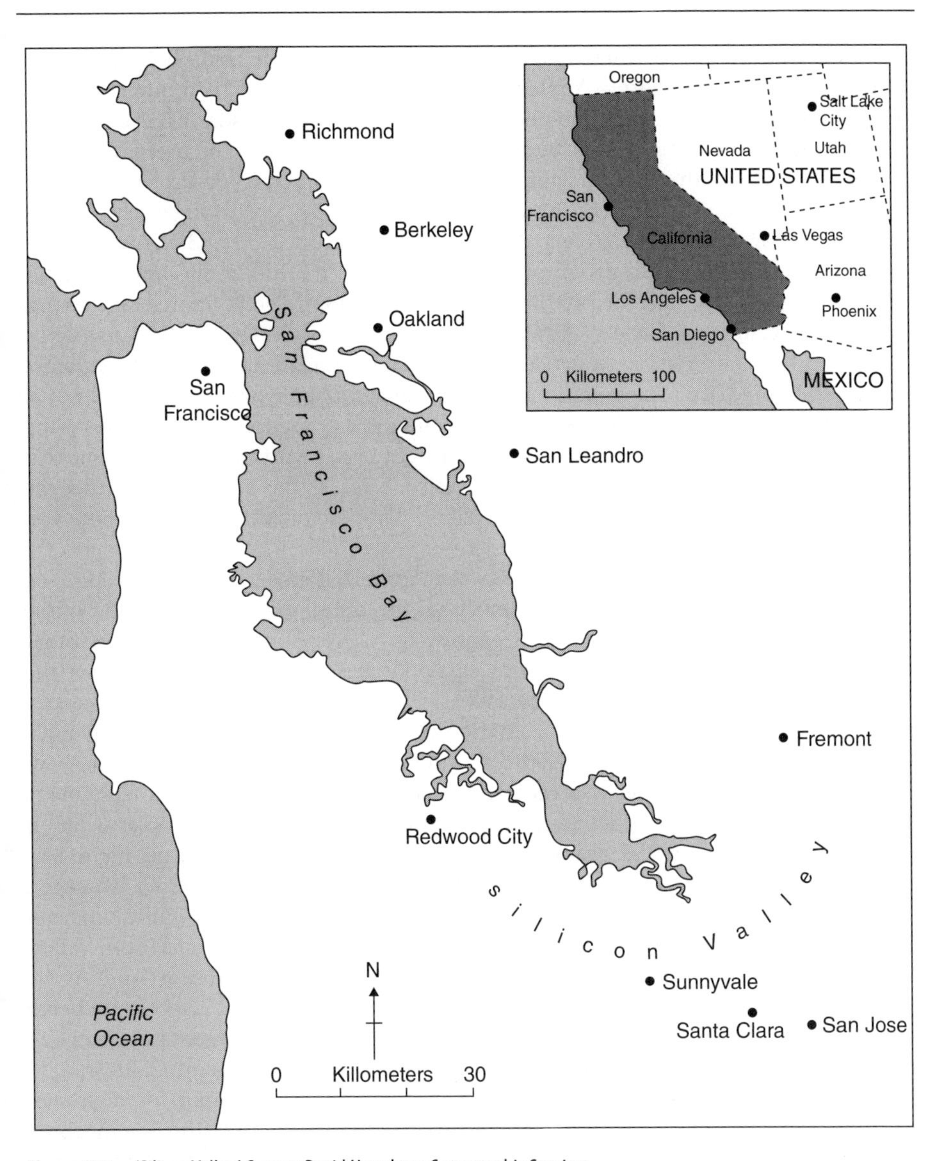

Figure 6.1 'Silicon Valley'. Source: David Houghton Cartographic Services.

in Calfornia' and these 'links with place provide the brand with a real spiritual home ... that people can connect to' (Beverland 2009: 149). Apple remains 'quintessential Silicon Valley ... like American cars and their associations with Detroit' (Brand Web Site Publisher and Editor, author's interview, 2013).

Rejecting 'names like Executek and Matrix Electronics' (Moritz 2009: 144), Apple Computer was the initial company and brand name selected by Steve Jobs: 'I was on one of my fruitarian diets. I had just come back from the apple farm. It sounded fun, spirited, and not intimidating. Apple took the edge off the word "computer". Plus, it would get us ahead of Atari in the phonebook' (quoted on Isaacson 2011: 63). Apple provided the name, visual identity and brand equity attributes of friendliness and quirkiness in which 'there was a whiff of counterculture, back-to-nature earthiness to it, yet nothing could be more American. And the two words together – Apple Computer – provided an amusing disjuncture' (Isaacson 2011: 63). Apple's early brand image symbolized its geographical associations: 'the bitten into apple is laden with rich suggestions, and the rainbow of the original logo also connotes the cultural blend that is California society' (Chevalier and Mazzalovo 2004: 33). Differentiation was integral to the brand's advertisers: 'Our whole strategic position is to enhance position for Apple. We have to add brand preference ... Apple has to be careful not to get down and dirty. An Apple means more than just a computer. Apple means plugging yourself into an energy source. We can't fight it out with the unwashed masses. We've got to show Apple is the brand of choice. Stick in the price bracket and you become a cheapo' (Maurice Goldman, Chiat Day advertising agency, quoted in Moritz 2009: 66–7).

Led by its market shaping innovations the Apple II Personal Computer (PC) in 1978 and the Macintosh (the 'Mac') in 1984, the brand experienced strong initial growth. Ease of use and 'simple and elegant design ... set Apple apart from the other machines, with their clunky grey metal cases' (Isaacson 2011: 73). Apple 'set the standard for the industry ... since the 1980s ... Apple II computers were beige as a neutral colour to fit well into office environments ... now its become synonymous with Silicon Valley style' (Brand Web Site Publisher and Editor, author's interview, 2013). Brand differentiation was central: 'we do not want to sell it as a technology machine. We want the product to have a personality and we want people to buy it because of its personality. We want to make it a cult product. We want people to buy it for its image as well as its utility' (Michael Murray, Apple Marketing Manager, quoted in Moritz 2009: 192). Articulating strong and valuable geographical associations in the emergent international centre of the high-tech industry in Silicon Valley, Apple's products were originated in their Californian and American heartland. At the time, '"Made in America" was a bigger deal back then in the 1980s ... jobs were being lost ... people wanted manufacturing in America' (Brand Commentator, author's interview, 2013). Apple Computer opened a new factory in Fremont, California, in 1984 to produce its Macintosh PC and it was lauded as a flagship for high-tech manufacturing' in America with Steve Jobs promoting the idea that 'this is a machine that is made in America' (quoted in Sanger 1984: 1). Echoing the vertical integration and mass marketing of automobile pioneer Henry Ford in Detroit in the early 1900s: 'Jobs spun an enticing vision of a computer for the masses, with a friendly interface, which would be churned out by the millions in an automated California factory ... the dream factory sucking in the California silicon components and turning out finished Macintoshes' (Isaacson 2011: 173).

Apple's initial production expansion faltered in the early 1980s. IBM entered the PC market with an open system based on Microsoft Windows software and Intel microprocessors that became the industry standard and enabled rapid capture of market share. The Windows and Intel combination encapsulated the economic and business model of 'Wintelism'. This model referred to the competitive shift from vertical control of markets by final assemblers towards more distributed market power across the value chain encompassing product architectures, components and software (Borrus and Zysman 1997). Intensified competition fomented a collapse in sales and net income reduction of 62% between 1981 and 1984, forcing Apple into crisis and precipitating the departure of Steve Jobs (Yoffie and Rossano 2012). Apple's management reacted by emulating its peers and relocating routinized production and assembly activity out of Silicon Valley to lower cost sites in the southern sunbelt states in America. The pilot factory in Cupertino was closed. Production was relocated to new plants in Carrollton, Texas, in 1980 (subsequently closed in 1985) and Fountain, Colorado, in 1985 (Prince and Plank 2012). Internationally, production extended to Europe for wider market access and reduced production costs. One of the first overseas Apple-owned plants was established in Cork, Ireland, in 1980 initially producing keyboards and employing nearly 1000 at its peak (Healy 2012). An Apple-owned plant for printed circuit boards was also established in Singapore (Prince and Plank 2012). Despite international production, the Apple brand managers retained the origination of 'Made in America' and there was 'no origination from the early international assembly plants ... there was never a "Made in Ireland" label' (Brand Web Site Publisher and Editor, author's interview, 2013).

Production grew and sales, market share and financial performance stabilized by 1990, underpinned by Apple's strongly differentiated brand and premium pricing. By the early 1990s, the originated distinctiveness of Apple and its geographical associations in Silicon Valley were under threat. Senior management 'tried to move Apple into the mainstream by becoming a low-cost producer of computers with mass-marketed appeal' by producing PCs 'designed to compete head-to-head with low-priced IBM clones' (Yoffie and Rossano 2012). The turn towards the mainstream and higher volume markets involved cost reduction, internationalization and the licensing of external manufacturers to produce Mac clones. Full-time employment was halved between 1995 and 1998 (Figure 6.2), and production concentrated at the Elk Grove, California, and Fountain, Colorado, plants (Prince and Plank 2012). The weakly differentiated and mass-market strategy for the Apple brand failed following static and then declining net sales and financial losses in 1996 (Figure 6.3). Further new senior management sought to move Apple out of the unprofitable battle for mass-market share and back to its higher margin differentiation and premium price model. Restructuring and cost cutting ensued, outsourcing internationally and rationalizing Apple-owned plants. The Cork plant lost 400 jobs following the shift of production of circuit boards to Singapore in 1992 (Healy 2012). The Fountain plant, Colorado, was sold to contract manufacturer SCI Systems to produce Macs for Apple. The circuit board plants in Singapore and Cork were sold to contract manufacturers too (Prince and Plank 2012). Without a new operating system to rival Microsoft Windows and with a widening and indistinct product line, the brand's coherence as a competitive commodity delivering commercial success unravelled. The differentiated meaning and value of Apple's geographical associations in Silicon Valley had been downplayed by its management's bid to compete in the spatial and temporal market setting of the

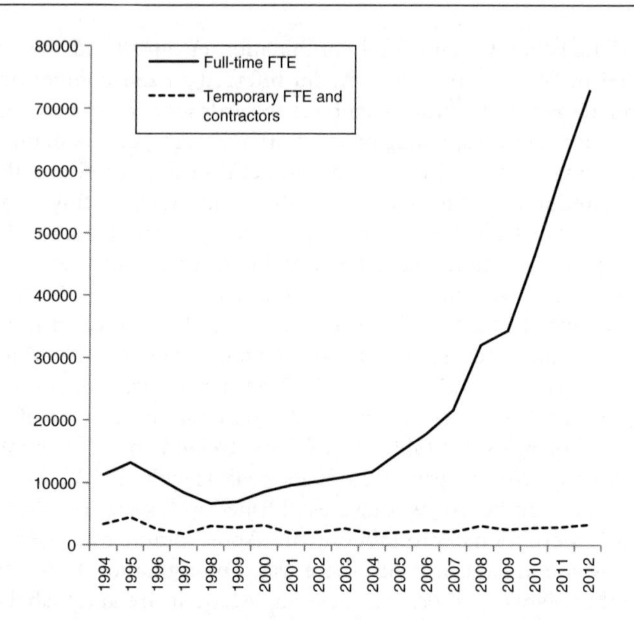

Figure 6.2 Employment in Apple by type, 1994–2012. Source: Calculated from Apple 10-K Annual Statements (1990–2012).

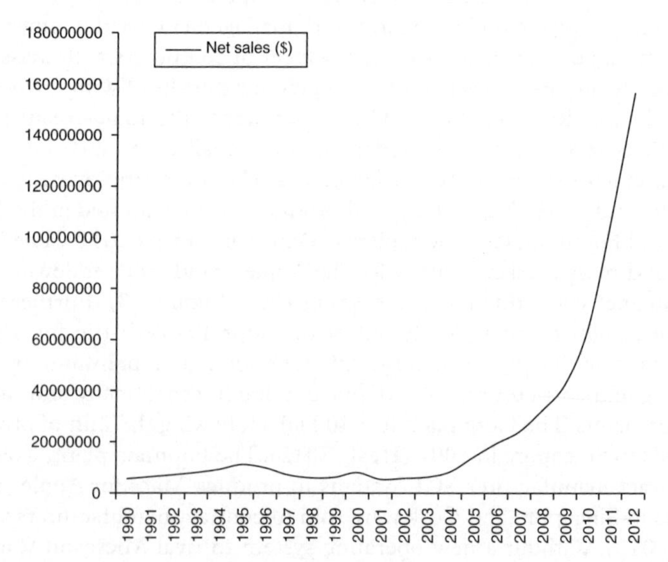

Figure 6.3 Apple Net sales by year, 1990–2012[a]. Source: Calculated from Apple 10-K Annual Statements (1990–2012).

Note: [a]Nominal prices.

mainstream. Worldwide market share declined to 3% and Apple posted a loss of $1.6 billion in 1996–1997: 'We were less than ninety days from being insolvent' (Steve Jobs quoted in Isaacson 2011: 339).

In a bid to save the company and salvage the brand, Steve Jobs returned as interim CEO. Apple's revitalization focused upon reconnecting the origination of the brand to Silicon Valley and reworking its meaning and value in the new market times and spaces of the late 1990s and 2000s. First, Steve Jobs crafted an innovative and globally oriented new strategy and business model for Apple as a 'mobile device company' (Steve Jobs, quoted in McLaughlin 2010). Critical was the reorientation of the Apple brand and the reframing rather than eclipse of the PC as the 'Digital Hub':

> the Macintosh had the real advantage for consumers who were becoming entrenched in a digital lifestyle, using digital cameras, portable music players, and digital camcorders, not to mention mobile phones. The Mac could be the preferred 'hub' to control, integrate and add value to these devices. Jobs viewed Apple's control of both hardware and software, one of the few remaining in the PC industry, as a unique strength. (Yoffie and Rossano 2012: 7)

Reflecting this shift from being only a PC company named 'Apple Computer', the company was renamed 'Apple Inc.' The identity and feel of the brand was refreshed with a minimalist monochrome logo to replace the rainbow-coloured image (Kahney 2002). Configuring a more 'global' outlook for the brand, Steve Jobs even interpreted a degree of 'global' convergence and homogenization in spatial and temporal market contexts in the digital age: 'It hit me that, for young people, this whole world is the same now. When we're making products, there is no such thing as a Turkish phone, or a music player that young people in Turkey would want that's different from one young people elsewhere would want. We're just one world now' (Steve Jobs, quoted in Isaacson 2011: 528).

Second, the actors involved renewed the culture and ethos at Apple to revitalize the brand's values and practices based upon focus, simplicity and elegance founded and originated in its radical innovation heritage geographically associated in Silicon Valley. In the brand's 'DNA' is the 'drive to simplify, to reduce ... you see it in the one button mouse, the one button iPhone, the graphical user interface ... making technology as accessible to a mass market ... so you didn't need a manual ... just pick it up and use it' (Brand Web Site Editor and Publisher, author's interview, 2013). The renewed approach was underpinned by increased levels of R&D spending – from $604 million in 1996 to $2872 million in 2012 (Apple 2012) – focused upon innovative and iconoclastic new Apple products with 'distinctive design that made a brand statement' (Isaacson 2011: 348). Design was elevated to an integral aspect of the brand's differentiation, revolutionized by now Senior Vice President of Industrial Design Jonathan Ive's interpretation of 'form follows function' design principles and design rather than engineering-led production strategy (von Borries *et al.* 2011). Apple's production management pioneered high value-added, refined and distinctively differentiated design capable of supporting premium pricing with low cost manufacture to 'design it at the highest quality ... while also ensuring that millions can be manufactured quickly and inexpensively enough to earn a significant profit' (Duhigg and Bradsher 2012: 3). This strategy depended upon innovative combination and packaging of generic and modular components under the Apple brand (Froud *et al.* 2012), the avoidance of differentiation for its own sake and a focus upon iteration through 'successive generations of the same product and improvement … stick with the design … the same basic machine' (Brand Web Site Editor and Publisher, author's interview, 2013). The new strategy yielded early commercial success with the 'first real coup' (Yoffie and Rossano 2012: 4) – the iMac – in 1998. The iMac featured colour options and translucent design, 'plug and play' ease of use and full interoperability with Microsoft software. iMacs were produced for the American market at Elk Grove, California. Reflecting the trend towards outsourced manufacture, iMacs were produced for Europe under contract by LG Electronics at a plant in Wales, UK (Prince and Plank 2012). The distinctive and differentiated meaning and value of the Apple brand were renewed by the iMac's success, setting the trajectory for further radical innovations in new mobile devices with the iPod, iPhone and iPad.

Third, Apple's production management shifted from a vertically integrated to disintegrated production model for the brand with a particular international reach and scale. Intensified rivalry and acute cost pressures rendered Apple uncompetitive against peers that had already outsourced manufacturing to lower cost contractors especially in East Asia (Mudambi 2008). In the Apple-owned factory in Elk Grove, California, in the early 1990s, for example, its iMac output was priced for sale at $1500. The build costs (excluding materials) were $22 compared to $6 in Singapore and $4.85 in Taiwan. The differentials resulted from inventory management and productivity differences rather than just labour costs due to their relatively low proportion of total costs (Duhigg and Bradsher 2012: 8). As the economics of production shifted, the global value chain for the Apple brand was reorganized. Functional specializations and integration amongst Apple's operations internationally were refined. The critical software developers remained in the Infinite Loop Campus in Cupertino, Santa Clara County. This location ensured the design, development and innovation of Apple's globally projected products and services were geographically associated and originated in Silicon Valley. Former manufacturing facilities in America and Europe were closed. The last manufacturing line in Elk Grove shut in 2004 with the loss of 235 jobs (Prince and Plank 2012). LG Electronics' subcontract to produce iMacs was cancelled following its inability to meet Apple's demand to reduce costs by $150–200 per unit (Smith 2000). Some former production facilities were shifted into consumer-oriented service functions, for example Elk Grove, California, and Cork, Ireland – the latter growing its employment to over 3000 in 2012 (Healy 2012). Diagnostics engineering was outsourced, for example to Singapore (Duhigg and Bradsher 2012). The company changed to a build-to-order system for manufacturing and distribution, mirroring Dell's then-successful business model. Assembly was outsourced to a small number of lead contractors for Apple branded products. Apple designers explain:

> that the move into China was organic. They were working with Japanese companies with factories in Taiwan. They wanted to move into metals and the supply chain was mostly in China. They started working with Alcoa [in America] but they weren't giving them the

> samples ... they weren't responsive enough ... Terry Gall at Foxconn would do anything to get the business and solve problems for Apple. (Brand Web Site Editor and Publisher, author's interview, 2013)

The lead contractors included Foxconn and Quanta Computer from Taiwan. Each operated plants across the world in Asia, Brazil, eastern Europe and Mexico and also supplied Amazon, Dell, Hewlett-Packard, Motorola, Nintendo, Nokia, Samsung and Sony (Duhigg and Bradsher 2012: 5). These top tier contractors managed the lower tiers of the supply chain, including specialized component suppliers integral to the function and branding of the new Apple products. These components included scratch-resistant 'Guerilla glass' for touch-screen devices from Corning Inc., Kentucky, and high performance semiconductors from Samsung's wafer fabrication plant in Austin, Texas.

Apple branded products now rely upon a sophisticated and internationalized supply chain (Table 6.1). Many suppliers are headquartered in the United States, Taiwan and Japan (Figure 6.4). Demonstrating the strong material geographical associations in Silicon Valley, 21 of the 45 US-headquartered suppliers are located in California. For Apple branded products such as the iPhone 'Though components differ between versions, all iPhones contain hundreds of parts, an estimated 90 per cent of which are manufactured abroad. Advanced semiconductors have come from Germany and Taiwan, memory from Korea and Japan, display panels and circuitry from Korea and Taiwan, chipsets from Europe and rare metals from Africa and Asia. And all of it is put together in China' (Duhigg and Bradsher 2012: 1). Mirroring the internationalization of Burberry's geographical associations of production in its 'national' origination (Chapter 5), the rationale behind the reorganization of production of the Apple brand is not a simple story of outsourcing for lower labour costs. The cost structure in the high-tech consumer electronics business is based upon specialized component parts and the integration of goods and services providers in overall supply chain management (Mudambi 2008). The main cost elements in an iPhone are display, camera and memory (Table 6.1). Integral to Apple's new production strategy is 'lean' supply chain reorganization, which:

> reduced the number of Apple's key suppliers from a hundred to twenty-four, forced them to cut better deals to keep the business, convinced many to locate next to Apple's plants, and closed ten of the company's 19 warehouses ... reducing the places where the inventory could pile up ... reduced inventory ... cut[ting] the production process for making an Apple computer from four months to two. All of this not only saved money, it also allowed each new computer to have the very latest components available. (Isaacson 2011: 361)

Even with the new outsourced production model, Apple's growth meant direct employment increased dramatically from its trough of 6558 in 1998 to 72 800 in 2012 (Figure 6.1). Apple now claims to support 307 250 jobs in America, including 50 250 directly at Apple – making it 22 550 employed directly by Apple outside America – and 257 000 jobs in other goods and services suppliers (Apple 2013). The lead assembly contractor's workforce has been estimated at 700 000 people who 'engineer, build and

Table 6.1 Preliminary Bill of Materials estimate for the 16 GB version of the iPhone 4, 2012[a]

Part description	*Sub-section*	*Cost (US$)*
Display	Display/camera	28.50
Flash	Memory	27.00
Electro-mechanics	Other	14.40
DRAM memory	Applications Processor	13.80
Baseband	Radio frequency	11.72
Applications processor	Applications Processor	10.75
Mechanicals	Other	10.80
Touch screen	Display/camera	10.00
Camera	Display/camera	9.75
Misc. RF components	Radio frequency	8.25
WiFi/BT	Connectivity	7.80
Battery	Battery	5.80
Misc. RF components	Other	5.50
Misc interface and sensor Components	Interface and sensors	3.80
Memory	Radio frequency	2.70
Gyroscope	Interface and sensors	2.60
Transceiver	Radio Frequency	2.33
Main PM device	Power management	2.03
Misc power management	Power management	1.90
GPS	Connectivity	1.75
Touchscreen Controller	Interface and sensors	1.23
Audio CODEC	Interface and sensors	1.15
Misc. connectivity	Connectivity	0.80
E-compass	Interface and sensors	0.70
Accelerometer	Interface and sensors	0.65
Misc. applications/ processor	Applications Processor	0.50
Misc. memory	Memory	0.30

Note: [a]Nominal prices.
Source: iSuppli (2013).

assemble iPads, iPhones and Apple's other products. But almost none of them work in the United States. Instead, they work for foreign companies in Asia, Europe and elsewhere, at factories that almost all electronics designers rely upon to build their wares' (Duhigg and Bradsher 2012: 1). Apple's role in geographically uneven development emerged following claims of exploitative working conditions, worker suicides, labour unrest and agitation for trade union representation at its major Taiwanese-owned subcontractor Foxconn's assembly plants in southern China in 2010 (Hille and

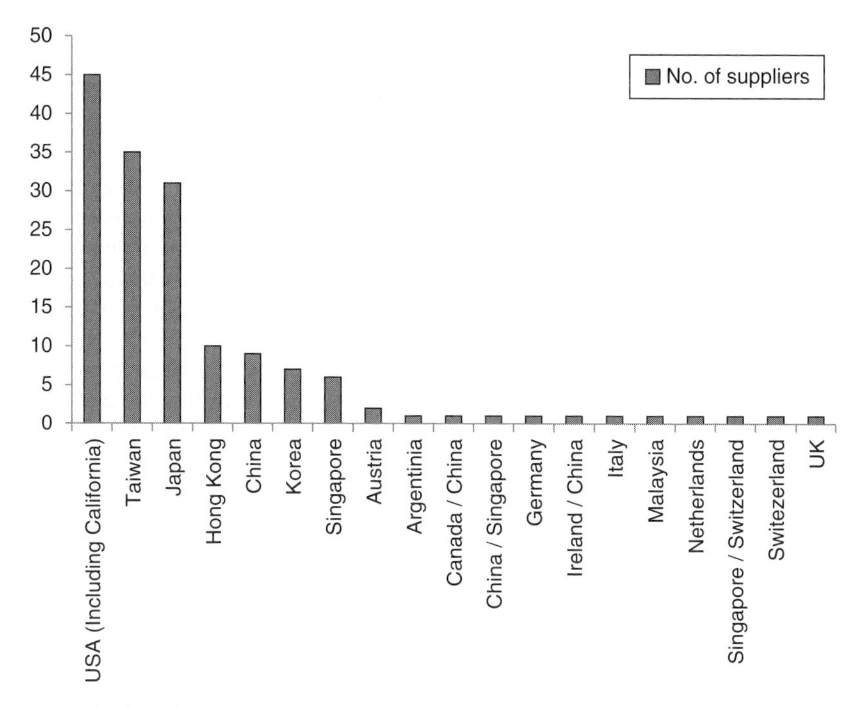

Figure 6.4 Apple supplier headquarters by country, 2012. Source: Calculated from Apple Supplier List (2013).

Jacob 2013). Foxconn was struggling to meet burgeoning international demand for Apple's latest iPhone and iPad products. Apple sought to ensure continuity of supply and contain the pressures of increased wage costs by encouraging its lead assembly contractor to shift production from its Shenzhen hub in Guangdong province to lower cost provinces in Anhui, Jianxi and Hubei in north and central China (Hille 2010).

International production outsourcing has meant originating Apple branded products in America has now become untenable:

> It isn't just that workers are cheaper abroad. Rather, Apple's executives believe the vast scale of overseas factories as well as the flexibility, diligence and industrial skills of foreign workers have so outpaced their American counterparts that 'Made in the U.S.A.' is no longer a viable option for most Apple products. (Duhigg and Bradsher 2012: 1)

A subtle change in the origination of the brand's flagship products has been undertaken to reflect the spatial organization of its different operations. The '*Designed by Apple in California. Assembled in China*' origination reassures globally savvy consumers that the innovation, design and engineering are undertaken at Apple's headquarters and R&D campus at Infinite Loop, Cupertino, California in America. The assembly location acknowledges the product is actually put together in China to enable Apple to reap wide profit margins from its differentiated premium price and cost efficient production. This origination fits Apple's strategy because 'they've looked at what it is that is actually really valuable and core to the company … design … bring out the point that's what we do … with the headquarters in Cupertino … we're responsible for this bit' (Technology Editor, author's interview, 2013). Apple's aim is to emphasize its Silicon Valley roots in the minds of consumers since the suite of 'i-products … coming from the ideation centre of the world … enables the price premium' (Professor of Strategic Management, author's interview, 2013). This origination has created misunderstanding amongst Apple consumers in America. Most recognized a large overseas manufacturing component but over 50% thought they were made partly in the United States, just under 20% thought entirely overseas or didn't know and just under 8% said entirely in America (Connelly 2012).

Apple's business model has attracted criticism following the global financial crisis from 2008 and national strategic economic concerns about industrial capacity and capability, investment, job creation, tax revenues and recovery in the domestic economy in America (see, for example, Froud *et al.* 2012). Charles Duhigg and Keith Bradsher (2012: 1) of the *New York Times* have termed this economic model 'the iPhone economy' or 'the iEconomy' (see also Lazonick 2010) – echoing the 'Wintelism' of the 1990s (Borrus and Zysman 1997). Amidst concern about the contracting middle class and their lack of confidence and spending hampering the recovery of the American economy, the high profile brand has been singled out: 'Apple's an example of why it's so hard to create middle-class jobs in the U.S. now. … If it's the pinnacle of capitalism, we should be worried' (Jared Bernstein, former economic adviser to the White House, quoted in Duhigg and Bradsher 2012: 2). Julie Froud *et al.* (2012) estimate that the relatively low proportion of labour costs in the total cost of the iPhone mean producing it in America would add an additional $65 per unit and only reduce Apple's profit per unit by several hundred dollars (see also Duhigg and Bradsher 2012). Managers at Apple argued that the speed, responsiveness, scale, technological sophistication, abundant engineering skills and flexibility at their lead contractor's plants in Asia cannot be matched in America. Maximizing profitability is Apple management's priority to generate funds for innovation investment given the aggressive technology-led competition in the high-tech business (Duhigg and Bradsher 2012). Reflecting the wave of 're-shoring' as the total cost dynamics of manufacturing tilt back towards the advanced industrial countries (Christopherson 2013) and returning to the origination of its history, the current CEO of Apple, Tim Cook, has since committed to transfer some manufacture of Apple computers from China to America. A facility owned by Quanta Computer Inc. from Taiwan at Fremont, Alameda County, in Silicon Valley, California, has been identified as the new assembly plant for a line of Macintosh computers (*The Economic Times* 2013).

Rather than emphasizing its ubiquitous and 'placeless' attributes, producing Apple as a 'global brand' (Technology Editor, author's interview, 2013) involves actors originating its meaning and value in the geographical associations in the particular place of Silicon Valley, California. This space and time has afforded a particular set of attributes, characteristics, ethos and outlook to the brand. Initial origination as 'Made in America' underpinned the brand's initial growth. Geographical associations to Silicon Valley were undermined in the context of competition, technological and market shifts inducing episodic attempts to capture share in homogenized mass markets, commercial and financial stress and decentralized and internationalized production. Crisis preceded revitalization and the re-origination of the brand back in the differentiated meaning and value of its founding in Silicon Valley but reworked for the changed and internationalized spatial and temporal market settings of the late 1990s and 2000s. The PC was reframed as the digital hub, focusing innovation upon mobile devices integrating hardware and software. Culture and ethos characteristic of Apple's radical innovation heritage in Silicon Valley were renewed with the return to focus, simplicity and design-led engineering. A new vertically disintegrated production model at the international scale reorganized Apple's global value chain through integration and management by lead contractors for cost efficient assembly. Origination shifted subtly from 'Made in America' to 'Designed by Apple in California. Assembled in China'. The geographically uneven development of Apple's operations internationally was reinforced, raising concerns about economic capacity, investment, job creation and recovery in the American economy as the 'iPhone' or 'iEconomy' emerges as an influential business and economic model.

Circulating the 'global' in Apple

Actors involved in originating Apple have tried to manage meaningful and valuable geographical associations in Silicon Valley in the relational networks of its internationalizing business and production model and promotion in worldwide spatial and temporal market settings. From Apple's founding, Steve Jobs 'understood, in ways that others did not, that the look and style of a product served to brand it' (Isaacson 2011: 193). For the brand, this image was 'quintessentially California style from the 1940s movement to the 1980s snow white movement' (Brand Web Site Editor and Publisher, author's interview, 2013). Differentiation emphasized how Apple would not 'stand a chance advertising with features and benefits and with RAMs and with charts and comparisons. The only chance we have of communicating is with feeling' (Steve Jobs quoted in Moritz 2009: 123). This brand-led focus has been central to the circulation strategies for Apple from its founding. Steve Jobs sought to 'create a consistent design language for all Apple products' to associate the brand's differentiated design with an internationally recognized and 'world-class' designer 'who would be for Apple what Dieter Rams was for Braun' (Isaacson 2011: 132). Dieter Rams' characteristically German modernist industrial designs pioneered the 'Less but better' (*Weniger aber besser*) principle. While aiming for a universally and internationally accessible modernism, this design language had its own geographical associations. It was used to originate meaning and value from the 'spare and functional philosophy' of the Bauhaus art movement from Germany, Japanese electronic company Sony's 'signature style and memorable product designs', and Apple's Silicon Valley-rooted ethos of rationality and functionality of 'clean lines and forms' articulated with 'capability for mass production' (Isaacson 2011: 126). Demonstrating the sometimes overlapping and messy geographical associations underpinning origination, the brand management evoked a particular sensibility. The German designer of Sony's Trinitron TVs Hartmut Esslinger was selected:

> Even though he was German, Esslinger proposed that there should be a 'born-in-America gene for Apple's DNA' that would produce a '*California global*' look, inspired by 'Hollywood and music, a bit of rebellion, and natural sex appeal'. His guiding principle was 'Form follows emotion', a play on the familiar maxim that form follows function … Jobs offered Esslinger a contract on the condition that he move to California … Esslinger's firm, frogdesign, opened in Palo Alto in mid-1983 with a $1.2 m annual contract to work for Apple, and from then on every Apple product has included the proud declaration 'Designed in California'. (Isaacson 2011: 133; emphasis added)

The 'California Global' origination melds geographical associations in the specific territorial scale of the state of California in America *and* the international relational networks and circuits connoted by the term 'global'. For Apple, with the global aspiration 'there's a Californification … they have that image … and they use that in their design … in terms of where do they come from … it is California not the US … there is a certain hipness, cachet to Silicon Valley … and they're its quintessential product' (Professor of Strategic Management, author's interview, 2013).

Early circulation of the Apple brand connected into networks of key trade commentators and journalists, building brand awareness through distinctive and memorable advertising campaigns. The original Victorian-style Apple logo adopted at establishment in the late 1970s was replaced. Seeking wider appeal, a new brand logo was developed, comprising an Apple shape with a bite taken out and six rainbow colour stripes. Circulation strategies changed incrementally during the episodes of Apple's growth and decline through the 1980s and 1990s. 'The Apple Marketing Philosophy' guided activities with its principles of empathy with consumer needs, focus and imputing the values and principles on which the brand is based (Mike Markkula, former chairman, quoted on Isaacson 2011: 78). The actors involved constantly tried to (re)frame the temporal and spatial market setting in which the Apple brand circulated. In the early 1980s PC segment, for example, competition was articulated as a 'two-way contest between the spunky and rebellious Apple and the establishment Goliath IBM, conveniently relegating to irrelevance companies such as Commodore, Tandy and Osborne that were doing just as well as Apple' (Isaacson 2011: 134).

Central to the late 1990s financial crisis was the Apple brand's loss of its distinctive differentiation and market position. Engaging in homogenized mass-market and low-cost competition, Apple undermined the meaning and value of the origination of its geographical associations with the revolutionary and innovative feel of Silicon Valley. Industry commentator Michael Gobe argued that 'the brand was pretty much gone. That's one of the reasons Apple's been rebranded – to rejuvenate the brand' (quoted in Kahney 2002: 1). A brand circulation priority for Steve Jobs on his return was 'to break away from Apple's tired, tarnished image. Jobs wanted Apple to be a cultural force' (Yoffie and Rossano 2012: 4). Critical to renewing and internationalizing the brand's meaning and value was the return to Apple's origination and reconnection with the revolutionary ethos geographically associated in Silicon Valley with a particular design-led revitalization of its differentiation.

Circulation strategies were changed in several ways. First, the new generation products, especially the mobile devices, were marketed emphasizing their compatibility and inter-operability with Microsoft Windows. A branded 'ecosystem' was constructed to draw consumers into the integrated world of Apple's hardware and software as well as accessories 'that ranged from fashionable cases to docking stations' (Yoffie and Rossano 2012: 8). Second, the product range was refocused on fewer key areas and rationalized. Third, the Digital Hub strategy and innovative mobile devices led modernization of Apple's brand image, repositioning Apple as 'evoking innovation and youth' characteristic of its geographical associations and origination in Silicon Valley (Steve Jobs, quoted in Isaacson 2011: 392). iPod advertising, for example, used distinctive and simple white silhouettes of the device, earphone leads and the user dancing on multiple coloured backgrounds to situate it in diverse individual social and cultural identities and musical tastes that 'work well across countries and cultures' (Hollis 2010: 180). It has even spawned a 'cult of iPod' (Kahney 2005). Fourth, launch events were used to 'turn the introduction of a new product into a moment of national excitement' (Isaacson 2011: 151), exploiting and amplifying Steve Jobs' guru-like standing and status amongst Apple users, consumers and the trade and general media as well as those uninitiated to the brand. Fifth, reprising its roots 'embodying the ethos of Silicon Valley' (Technology Editor, author's interview, 2013), Apple's innovative and revolutionary feel was revisited in promoting the brand 'as a hip alternative to other computer brands' and advertising was placed beyond the computer trade and user press in 'popular and fashion magazines' (Yoffie and Rossano 2012: 4). Last, the Apple web site became the focus of interaction and order, sales and distribution channel. This circulation strategy secured Apple's turnaround in the late 1990s, providing 'a way to reclaim the brand' (Lee Clow, creative director, Chiat/Day, quoted on Isaacson 2011: 143). It underpinned Apple's transformation into 'one of the world's most iconic brands ... everyone recognises the Apple logo ... even the most impoverished peasant in China"' (Brand Web Site Editor and Publisher, author's interview, 2013). Critical to the brand's circulation was the (re)construction of Apple by the participant actors as more than just a product with functional and use value but as a wider concept and symbol:

> Apple was one of the great brands of the world, probably in the top five based on emotional appeal, but they needed to remind folks what was distinctive about it. So they wanted a brand image campaign, not a set of advertisements featuring products. It was designed to celebrate not what the computers could do, but what creative people could do with computers. (Isaacson 2011: 328)

Last, the distribution system for Apple branded goods and services was changed to direct supply through new Apple stores and the Apple web site – supported by an integrated production, logistics and delivery network – and indirectly through national chains rather than smaller outlets.

Apple's international – even 'global' – circulation, high profile and commercial success has attracted contestation from actors who criticize the brand as embodying the socially, spatially and environmentally unequal and unsustainable nature of contemporary capitalism (see, for example, Klein 2010). Protest, including subverting the meaning of adverts or 'adbusting', has targeted the brand as a symbol of unsustainable global production and consumption patterns (Adbusters 2012). Apple's 'Think Different' advertising campaign images have been recirculated with altered images and messages, including Stalin with the slogan 'Think Really Different' and the Apple logo morphed into an image of a human skull with the strapline 'Think Doomed' (Klein 2010). Concerns about poor terms and conditions, low wages, long working hours and lack of independent trade union representation have been focused upon Apple's lead contractors. FoxConn's 230 000 employees typically work 12 hour shifts up to 6 days per week and earn less than $17 per day (Duhigg and Bradsher 2012: 5). The workforce is housed in company dormitories within so-called 'Foxconn City' in Shenzhen, China, and the company provides the social infrastructure of restaurants, basic health services and security. Such practices have stimulated dissent about labour intensification, control, domination and exploitation amongst campaign groups. The Apple's brand's circulation strategy has been subverted by the adbusting community with their 'iSlave' image, which has been used to highlight the plight of the workforce, raise awareness amongst consumers and agitate for remedial action. In a sign of the vulnerability of brands to undesirable geographical associations that risk tainting their meaning and value, the media attention and public outcry led Apple to enhance its supplier audit activities involving the Fair Labour Association to monitor labour conditions and practices (Hille and Jacob 2013).

Integral to circulating the 'global' in Apple are the tensions and accommodations between the relational networks of its increasingly international promotion and reach and the meaning and value of the brand rooted in its geographical associations in the particular place of Silicon Valley. The origination of 'California Global' geographically associated Apple with the territory of California – and specifically Silicon Valley – *and* the relational networks and circuits of the global. Some commentators emphasize how this makes Apple a 'non-geographical brand ... any consumer in any country in the world would see it as Apple ... successfully generating buzz and creating cultural connection that transcends Silicon Valley' (Brand Commentator, author's interview, 2013). Others claim that 'We may not have lived in Cupertino or be American, but we can relate to the creative spirit and history of the computing industry through using an Apple' (Beverland 2009: 150). Circulation techniques and practices were utilized to articulate and represent Apple's differentiation through an array of channels. Weakened distinctiveness presaged financial crisis as the brand competed on low prices in the homogenized mass market. Revitalization of the tarnished meaning and value of the brand from the late 1990s involved renewing Apple's origination and historical geographical associations with the innovative and revolutionary ethos and 'DNA' of the brand from Silicon Valley. More open marketing approaches, dramatic launch events and publicity, the emphasis on Apple's integrated Digital Hub 'ecosystem' and the lead role of new mobile devices connoting innovation and youth in the brand were utilized by the participant actors to shape its emotional appeal in the wider culture. Apple has developed global reach and status through the international marketing and availability of its products and services on-line and through its extensive network of distinctive design-led outlets and concessions. The brand's increasingly high and international profile attracted contest and subversion from activists and dissidents as a symbol of culturally, ecologically, economically, politically and socially unequal and unsustainable contemporary capitalism.

Consuming the 'global' in Apple

The Apple brand's initial commercial success in promoting purchase and consumption was firmly originated in its Silicon Valley hearth. Apple's revolutionary and innovative products and services with distinctive and differentiated design supported premium pricing and high profit margins. Throughout the technology business history:

> Apple was the only one with distinctive Silicon Valley roots ... the other OEMs weren't. Dell was from Texas, Compaq from Houston, Texas, and Microsoft from Seattle not Silicon Valley. Only HP [Hewlett Packard] was the product and engineering driven focused company from Silicon Valley. (Brand Web Site Editor and Publisher, author's interview, 2013)

Contrasting the world of cheap and functional PCs, the Apple brand's distinctive differentiation focused upon closely reflecting its consumers' emotional and functional needs and personalities. Apple's distinctive design image and personality has attracted and mobilized an enduring brand community of users and consumers with a cultural-economy of distinctive mores, practices and styles – what Leander Kahney (2006) calls the 'Cult of Mac' (see also Belk and Tumbat 2005; Moritz 2009). Distinctive differentiation built up consumer loyalty enabling Apple to sustain its premium pricing strategy and high levels of profitability (Yoffie and Rossano 2012). Despite the growth of devotee consumers, the late 1990s financial problems were generated when Apple management chased share in mass consumer markets with lower prices, relatively less profitable volume sales and the third party licensing of Mac clones. As Simon Spence (2002a: 3) put it, Apple 'lost touch with its origins'.

Renewal of Apple's consumption momentum followed the Digital Hub strategy, returning the brand to the radical inventiveness characteristic of its origination and founding in Silicon Valley. As net sales of Apple products grew spectacularly from the 1990s (Figure 6.3), consumption of the brand internationalized with increases of 1400% in America and the Americas, 1700% in Europe and 3600% in the rest of the world (Table 6.2). In a version of Say's Law, Steve Jobs and the staff at Apple 'developed not merely modest product advances based on focus groups, but whole new devices and services that consumers did not yet know they needed' (Isaacson 2011: xix). The new devices generated dramatic sales growth internationally and leading positions in their

Table 6.2 Change in net sales by operating segment ($'000 s), 1992–2012[a]

Operating segment	*Net sales 1992*	*Net sales 2012*	*Change 1992–2012*	*% Change 1992–2012*
Americas[b]	3 885 042	57 512 000	+53 626 958	1380
Europe	201 784	36 323 000	+34 305 160	1700
Other countries[c]	118 366	43 845 000	+42 661 340	3604

Note: [a]Nominal prices; [b]'United States' in 1992; [c]'Pacific' and 'Other countries' in 1992, includes 'Asia-Pacific' and 'Japan' in 2012.
Source: Calculated from Apple 10-K Statements (1992–2012).

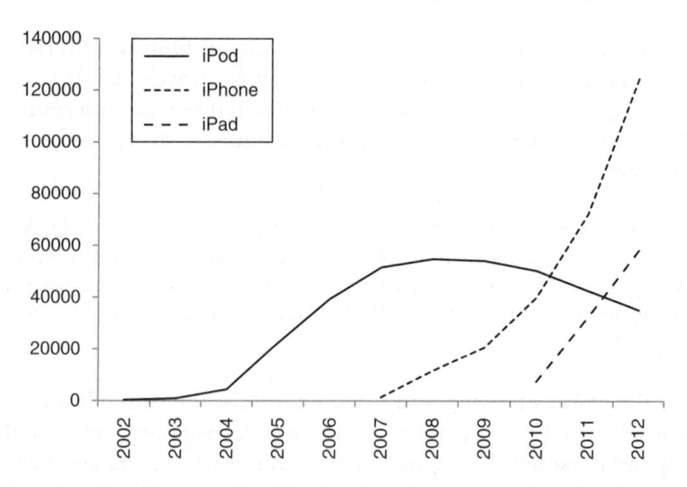

Figure 6.5 Unit sales of iPod, iPhone and iPad ('000 s), 2002–2012. Source: Calculated form Apple 10-K Statements (2002–2012).

markets (Figure 6.5). Some analysts have claimed that 'Jobs and Apple have always seen themselves as architects … not really radical innovators … they show people new ways of putting things together and designing things that are beautiful … three elements to it: architecture, design and brand/image' (Professor of Strategic Management, author's interview, 2013). Drawing upon increased levels of expenditure in R&D, a proprietary set of devices and applications were developed with the Apple brand's distinctive i- prefix branding: 'The "i", Jobs later explained, was to emphasize that the devices would be seamlessly integrated with the Internet' (Isaacson 2011: 338).

Although a late entrant to the digital music player segment in 2001, the iPod disrupted existing market contexts with its distinctive Apple branded design, simplicity, ease of use, user friendly interfaces, larger memory, longer battery life and easier handling of content than the competition. Crucial to iPod's innovation and differentiation was its design and connection to the on-line iTunes digital music store:

> To make the iPod really easy to use … we needed to limit what the device itself would do. Instead we put the functionality in iTunes on the computer … we made it so you couldn't make playlists using the device. You made playlists on iTunes, and then you synced the

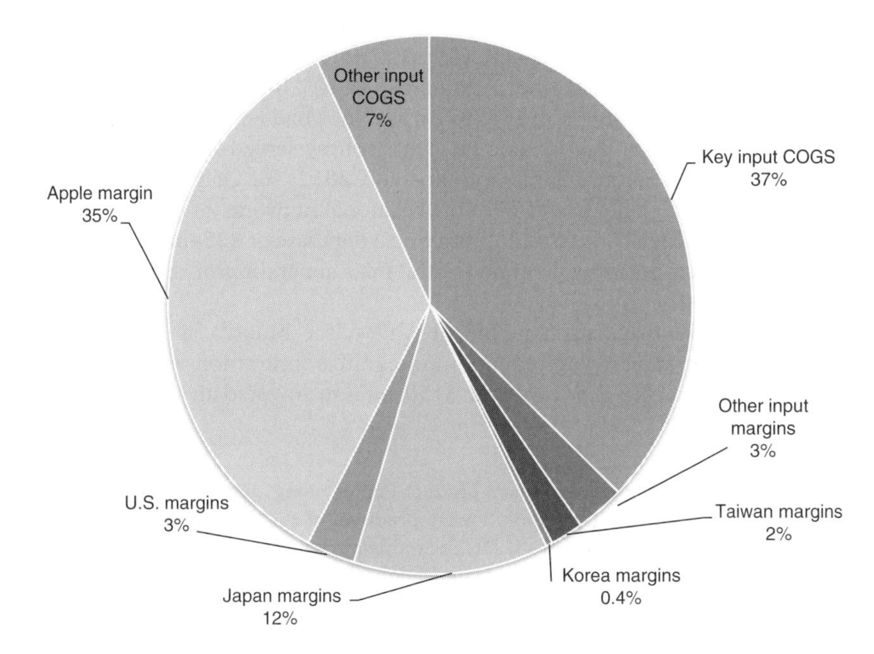

Figure 6.6 Value capture in a Video iPod (30 G) as percentage of wholesale price. Source: Adapted from Dedrick *et al.* (2009: 19).
Note: COGS is cost of goods sold including purchased inputs and direct labour.

> device. ... So by owning the iTunes software and the iPod device, that allowed us to make the computer and device work together, and it allowed us to put the complexity in the right place. (Steve Jobs, quoted in Isaacson 2011: 389)

The distinctive branded packaging meant 'When you took an iPod out of the box, it was so beautiful that it seemed to glow, and it made all other music players look as if they had been designed and manufactured in Uzbekistan' (Steve Jobs, quoted in Isaacson 2011: 393). The device generated linked demand for Apple desk and laptop PCs as the Digital Hubs, promoting complementary sales across the branded range. Branded differentiation supported Apple's highly profitable premium pricing strategy, enabling a comparatively high level of value capture and margin (Dedrick *et al.* 2009) (Figure 6.6).

iTunes bore the hallmarks of Apple's differentiated brand: distinctive and signature design, ease of use, and simplicity. And it created switching costs by locking users into Apple's tightly integrated ecosystem of branded products and services (Froud *et al.* 2012). In the context of Apple's 'global' aspirations, iTunes has 'localised stores in different countries ... Apple would like one store and have everyone access it ... brand association in buying even digital goods from the States and California ... via a window on a machine and server capacity from wherever' (Brand Web Site Editor and Publisher, author's interview, 2013). Demonstrating the compromises between territorial jurisdictional and relational networked spaces, record company licensing issues and EU regulators prevent actors involved with Apple creating a single integrated store even in on-line space. iTunes is not especially profitable for Apple. Sales revenue breakdowns have been estimated at 70% for the music label owner, 20% to credit card processing and 10% to Apple: 'Jobs had created a razor-and-blade business, only in reverse: the variable element (songs) served as a loss leader for a profit-driving durable good' (Yoffie and Rossano 2012: 9). Of greater longer-term strategic importance for the brand were the relational networks of consumers worldwide established as 'the iTunes Store … built up a database of 225 million active users by June 2011, which positioned Apple for the next generation of digital commerce' (Isaacson 2011: 410).

Reflecting Apple's origination in Silicon Valley, the brand's late entry into the smartphone market sought differentiation, integrative innovation and revolutionary representation. Following 2.5 years and $150 million invested in development, the iPhone was launched:

> Every once in a while a revolutionary product comes along that changes everything. Today, we're introducing three revolutionary products of this class. The first one is a widescreen iPod with touch controls. The second is a revolutionary mobile phone. And the third is a breakthrough Internet communications device. … These are not three separate devices, this is one device, and we are calling it iPhone. (Steve Jobs quoted in Yoffie and Rossano 2012: 9)

Apple management's attempts to control iPhone sales and consumption had to evolve to secure control over distribution and consumption of the brand. The initial model only allowed iPhones to work with a single telecommunications services provider contracted with a revenue-share agreement. But consumers purchased iPhones from unauthorized 'grey market' resellers and hacked them to enable use on a wider range of mobile networks (Yoffie and Rossano 2012). Given the potential loss of control over the brand and sales revenue, the distribution model was shifted to multiple rather than single carriers contracted to provide a subsidy to Apple per phone. iPhone sales growth propelled Apple into a rivalry with Samsung for share of the expanding smartphone market. The brand retained its higher level of profitability and iPhone provided over 40% of Apple's total revenue in 2012 (Yoffie and Rossano 2012).

Software applications – or 'Apps' in the lexicon of the digital age – provided a further branding and distribution opportunity for Apple. The emergence of Apps enabled the establishment of a relational and networked 'brandspace' on-line. Introduced as part of iTunes, the Apple App Store provided an accessible and easy to use branded platform for the downloading of Apps direct to especially Apple devices. Apple's brand control was exercised through its prior approval of all applications provided in the App Store. The economics of Apps were based on encouraging widespread take-up and advertising links because many were free and 'crowd-sourced' from an international commercial and renegade development community, and only a small number became 'viral' hits. The App Store registered over 4 billion downloads worldwide in its first 18 months and 585 000 different applications were available by 2012; Apple contracted to retain 30% of the developer's App sales and generated $6.3 billion in revenues from the sales of music, books and applications in 2011 (Yoffie and Rossano 2012).

Again invoking Apple's particular version of Say's Law in relation to consumers, the iPad disrupted the existing tablet computer category. The tablet market was a small niche with little signs of growth. Apple's late market entry built upon its capability and reputation for design-led innovation, rooted in the origination of the brand in Silicon Valley, to rework an existing product type and its meaning and value in an international spatial and temporal market context. Characteristic Apple brand attributes marked iPad's design and performance: ease of use, software and hardware integration, and the 'wow' factor of its design. Differentiated premium pricing and high volume sales, generated a gross margin of 25% on the entry level model by using Apple's own Central Processing Unit and leveraging its scale economies in purchasing and manufacture (Yoffie and Rossano 2012). By 2012, iPad sales reached 55 million and generated $35 billion in sales while over 20 competitor tablets were launched by 2011 including devices from Samsung, Amazon and Microsoft (Yoffie and Rossano 2012).

Despite Dell Computer adopting a direct sales channel and Gateway facing financial losses on suburban retail outlets (Isaacson 2011), Apple management changed the brand's retail distribution strategy from indirect via wholesalers to direct via Apple owned and branded stores in the early 2000s. With the growing international reach of the Apple brand, 'For anyone outside Cupertino, there were few examples of a firm presence to point to and say "There's Apple"' (Spence 2002b: 2). Extending its origination in time and space, Apple's branded retail stores 'gave this California-centric company beachheads across the globe' (Lashinsky 2012: 152). Balancing its Silicon Valley origination and global ambition, Apple wanted 'to sell products that are home grown all over the world ... [it] didn't want to localise ... only the language and the OS [Operating System]' (Brand Web Site Editor and Publisher, author's interview 2013). The first of four trial stores opened in McLean, Virginia, in 2001. Retail strategy changed for several reasons. First, Apple's management wanted to integrate and secure control over the user experience of purchasing branded goods and services rather than relying upon third parties:

> Industry sales were shifting from local computer speciality shops to megachains and big box stores, where most clerks had neither the knowledge nor the incentive to explain the distinctive nature of Apple products. ... Other computers were pretty generic, but Apple's had innovative features and a higher price tag ... [Steve Jobs] didn't want an iMac to sit on a shelf between a Dell and a Compaq while an uninformed clerk recited the specs of each. (Isaacson 2011: 368)

Second, Apple stores delivered brand awareness and profile, encouraging brand literacy and attracting footfall to recruit new consumers. Geographical association in the origination of the brand was central to their design as 'computers, software and consumer electronic devices were displayed in an atmosphere that was like a breath of fresh, California coastal air' (Moritz 2009: 338). But, this brand space was distinctive so 'you know you're in an Apple store regardless of location' (Lashinsky 2012: 152). Each store in every country followed the differentiated Apple design aesthetic adapted

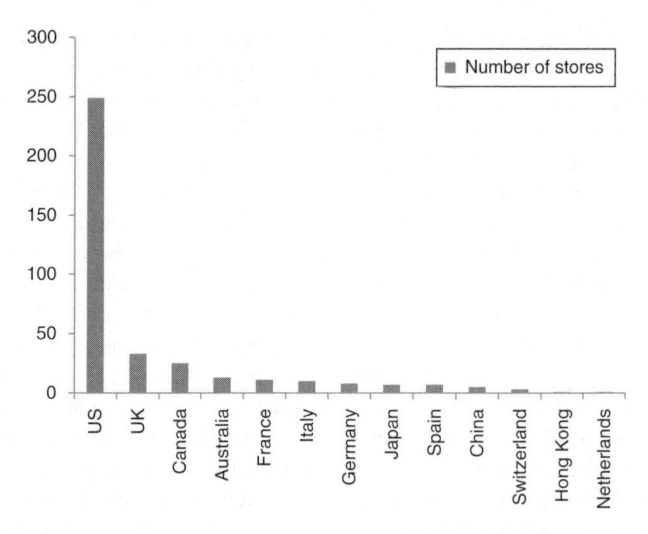

Figure 6.7 Apple stores by country, 2012. Source: Calculated from Apple 10-K Statements.

to fit into their local context: 'the stores have a distinctive look where they match their surroundings ... but inside it's the same experience, same wood tables, same blue t-shirts, same vibe, same experience ... it works, its friendly, cool' (Brand Commentator, author's interview, 2013). Indeed, the meaning and value of the brand and its distinctive differentiation meant 'it would take the relationship between retailing and brand image to a new level. It would also ensure that the consumers did not see Apple computers as merely a commodity product like Dell or Compaq' (Isaacson 2011: 374).

Third, Apple retail stores provided brand prescence in key sites 'in areas with lots of foot traffic, no matter how expensive' (Isaacson 2011: 369). In common with Burberry's flagship retail stores (Chapter 5):

> Many of Jobs's passions came together for Manhattan's Fifth Avenue store, which opened in 2006: a cube, a signature staircase, glass, and making a maximum statement through minimalism. ... Open 24/7, it vindicated the strategy of finding signature high-traffic locations by attracting fifty thousand visitors a week during its first year. (Isaacson 2011: 376)

Apple's retail outlets grew to 326 across 13 countries by 2011 (Figure 6.7). Despite its global aspirations, most stores are located in America and have 'got to saturation in the US ... everywhere you go ... so the new ones are in Europe, Australia and elsewhere" (Brand Commentator, author's interview, 2013). Key city locations in emerging and growing markets are especially important in building brand awareness, for example the IFC Mall in Pudong, Shanghai, China. In 2004, the stores were averaging 5400 visitors per week – compared to 250 per week at Gateway – average annual revenue per store was $34 million, and total net sales were $9.8 billion in 2010 (Isaacson 2011: 376).

Tensions and potential disruptions continue to shape the consumption of the 'global' in Apple branded products and services internationally. First, Apple's emergence as a pervasive international and in some commentator's views 'ubiquitous' brand (Copulsky 2011: 14) has generated disquiet amongst its longstanding and strong brand community (Muñiz and O'Guinn 2001). The Cult of Mac web site, for example, reflects the anti- and pro-Apple voices, the community's continued identification with its ever-expanding branded products and services, injustice in competitors mimicking attributes and features, jealousy about the dramatic sales growth embedding the brand in the mass market, and the effects of the brand's powerful advertising (see, for example, Elgan 2012). Second, Apple faces intense and shifting competition in fluid and internationalized technology-based spatial and temporal market settings. New innovations such as the open and free Android platform – contrasting with Apple's more closed and higher priced version – have underpinned rival systems and competitors' growth. Third, the business model for the Apple brand is based upon a continuous search for monopoly rents through tight integration, high switching costs and design-led differentiation. Despite the extension of Apple's international reach and consumption, the brand remains originated and managed from the company's HQ since 'Cupertino now has absolute say over the silicon, device, operating system, App Store, and payment system' (Jon Fortt, *Fortune*, quoted in Isaacson 2011: 496–7). Yet, what Julie Froud *et al.* (2012: 20–1) call a 'jackpot business model' is 'inherently fragile because it is always dependent upon the next hit product'. In the context of long-term transitions towards more open innovation systems (Chesbrough 2003), Apple's relatively closed world and its spatial concentration in Cupertino place a heavy emphasis upon its geographical associations in Silicon Valley to sustain its leadership in harnessing disruptive innovations in fluid and fast moving market contexts. Commercial failures with the Newton PDA, Mac Mini entry-level desk top and Apple TV demonstrate the risks involved. Disruptive innovations and ongoing refinement and upgrading of products and services cannibalize existing offerings when new devices integrate and outperform the older versions. iPhone handles music better than iPod, for example, and iPad Mini displaces iPad sales. Last, Apple's international consumption is complicating attempts to refine the geographical associations in the brand's origination. Evidence of how the originations matter is ambiguous:

> Do Apple consumers care where it is made? ... yes and no ... if they didn't care then there would have been no uproar over Foxconn ... and no because it has never really affected sales ... it has become accepted fact of the world ... 'Made in China' ... and south east Asia ... not unique to Apple products ... as a consumer you can't do much about it ... if you wanted to boycott you would not have much left to buy. (Technology Editor, author's interview, 2013)

The brand managers are addressing such disjunctures through subtly different articulations of origination in different market times and spaces. iPhones on sale in China carry only the '*Designed by Apple in California*' labelling (Nussbaum 2009). The '*Assembled in China*' part has been removed because of its negative connotations amongst aspirant middle class Chinese consumers who would 'prefer to buy brands made outside China as they think they would be better quality' (Chinese consumer, author's interview, 2013).

Consuming the 'global' in Apple is strongly entwined with the brand's origination in the geographical associations of the particular place of Silicon Valley. Early sales growth relied upon close connections between the distinctive differentiation of the brand and the rise and reputation of the high-tech centre of Silicon Valley within the American market. Recovering from IBM's disruption of the PC market in the early 1980s by focusing upon specific niches, the closed and tightly integrated system characteristic of the Apple brand was honed and supported early internationalization beyond the American market. Distinctive differentiation underpinned growing brand awareness and loyalty, achieving cult status in some circles. Consumption momentum was lost and financial crisis fomented as Apple strayed from the characteristic attributes of its Silicon Valley roots with a foray into low price competition in the more homogenized mass market. Revitalization from the late 1990s was constructed through renewing the differentiated, revolutionary and innovative ethos and values of the brand geographically associated with its Silicon Valley origination. Introducing a new strategy and suite of integrated products and services, the actors involved rearticulated the lustre and youthfulness of the Apple brand, relating and embedding its values in the wider cultural economy. Retailing was shifted from indirect to direct channels through the establishment of Apple branded stores internationally, situated in key locations to support the high profile of the brand, build awareness and literacy, and generate complementary sales of related products and services. Tensions and accommodations in Apple's international consumption reach comprised its relationships to its existing core consumers, intensified competition and reliance upon its relatively closed and integrated innovation model geographically centred where the brand is originated in Cupertino, Silicon Valley.

Regulating the 'global' in Apple

As the company's most important intangible asset – valued at $21,143 million in 2009 and ranking 19th worldwide (Interbrand 2010) – and its growing international reach through the 2000s, the Apple brand is a source of meaning and value tightly regulated through ownership and intellectual property rights. In 2013, Apple Inc. had 5266 utility and 984 design patents registered, 185 trademarks and 78 service marks for Apple branded training, educational and other support services (US Patent and Trademark Office, Personal Communication, 2013). In the particular spatial and temporal market settings of the international high-technology business, creating and protecting intellectual property has been critical from the founding of Apple in Silicon Valley in the late 1970s:

> From the earliest days at Apple, I realized that we thrived when we created intellectual property. If people copied or stole our software, we'd be out of business. If it weren't protected, there'd be no incentive for us to make new software or product designs. If intellectual property begins to disappear, creative companies will disappear or never get started. (Steve Jobs, quoted in Isaacson 2011: 396)

By 2011, the web of regulatory protection integral to the meaning and value of the Apple brand reached widely and deeply across its branded goods and services, including design, colour, technologies and packaging. As actors sought to protect both the 'global' in Apple and its origination in Silicon Valley, the importance of regulation was accentuated by the international reach and sales footprint of the brand and its dramatic growth in sales.

The strong need for R&D investment in the technology intensive business of PCs meant Apple's co-founders framed the brand as a financial asset to attract further injections of capital from the outset of the company in the late 1970s. In the particular context of the American variegation of stock-market oriented capitalism (Peck and Theodore 2007) and Silicon Valley's rich capital market infrastructure (Saxenian 1996), Apple launched its first Initial Public Offering (IPO) in 1980, only 4 years after the founding of the company. The public listing put a price on the company and capitalized financially on its initial growth:

> Apple Computer Co. in January 1977, they valued it at $5,309. Less than four years later they decided it was time to take it public. It would become the most oversubscribed initial public offering since that of Ford Motors in 1956. By the end of December 1980, Apple would be valued at $1.79bn. (Isaacson 2011: 102)

Flotation tied Apple into the capital markets early on in its development and its share ownership base changed from venture capital and regional companies in Silicon Valley to mainstream and internationalized capital market institutions (Moritz 2009). Current major shareholders comprise major institutional investors such as Vanguard Group, FMR, the State Street Corporation, Jupiter Asset Management and Barclays Global Investors UK Holdings (Apple 2012). The historical origination of Apple in the particular hippy and West Coast techno-culture of Silicon Valley combined with its rollercoaster financial performance and recent rapid growth and size as a financial entity framed a distinct and at times fractious relationship between Apple and its investors. From the outset and at its first signs of financial turbulence in the mid-1980s, 'East Coast stockholders always worried about California flakes running the company' (Tech stock newsletter editor, quoted in Isaacson 2011: 217). Apple's share price increased when Steve Jobs was forced out in 1985.

Following initial growth, Apple endured financial fluctuations in the mid-1980s, recovering from the late 1980s and into the mid-1990s before plunging into a financial crisis in the late 1990s (Figures 6.8 and 6.9). Apple's management struggled to maintain the coherence of the brand as a differentiated and competitive commodity in its particular spatial and temporal market contexts. The brand's distinctive design-led differentiation was under pressure from financial imperatives and cost cutting pressures. At the time, Apple almost lost the design guru Jonathan Ive central to its eventual recovery who was 'sick of the company's focus on profit maximisation rather than product design' because:

> there wasn't that feeling of putting care into a product, because we were trying to maximise the money we made. All they wanted from us designers was a model of what

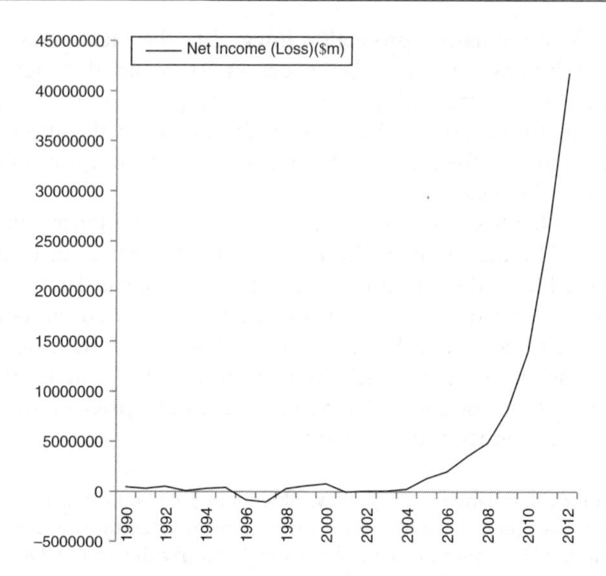

Figure 6.8 Net income by year, 1990–2012[a]. Source: Calculated form Apple 10-K Statements (1990–2012). Note: [a]Nominal prices.

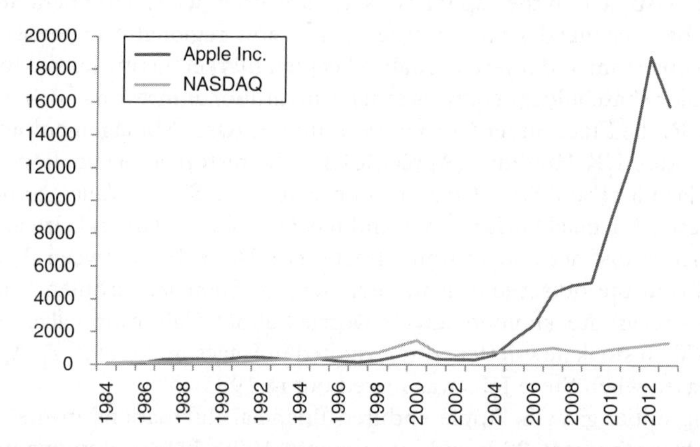

Figure 6.9 Apple share price and NASDAQ index, 1984–2013 (1984=100)[a]. Source: Calculated from Apple share price and NASDAQ data; http://uk.finance.yahoo.com/q/hp?s=AAPL,http://uk.finance.yahoo.com/q/hp?s=%5EIXIC. Note: [a]Annual average from monthly adjusted closing price. Nominal prices.

> something was supposed to look like on the outside, and then engineers would make it as cheap as possible. (Jony Ive, quoted in Isaacson 2011: 341–2)

The brand's market share fell from its 16% peak in the late 1980s to 4% in 1996. Sales revenues collapsed and Apple 'lost $1 billion, and the stock price, which had been $70 in 1991, fell to $14, even as the tech bubble was pushing other stocks into the stratosphere' (Isaacson 2011: 296–7) (Figure 6.10). The brand's financial plight even encouraged failed attempts by management to sell the company to Sun, IBM and Hewlett Packard (Isaacson 2011). Apple was on the brink: 'Apple Computer, Silicon Valley's paragon of dysfunctional management and fumbled techno-dreams, is back in crisis mode, scrambling lugubriously in slow motion to deal with imploding sales, a floundering technology strategy, and a haemorrhaging brand name' (Brent Schendler, *Fortune* technology report, quoted in Isaacson 2011: 311). Competitors were scathing about Apple's plight, as Michael Dell, CEO, Dell Computers, put it: 'What would I do? I'd shut it down and give the money back to shareholders' (quoted in Moritz 2009: 335).

The commercial and financial turnaround for Apple began in the late 1990s following the return of Steve Jobs, the development of the Digital Hub strategy, the reorientation towards mobile devices, growing international reach, and the design-led revitalization of the distinctive differentiation of the Apple brand. The re-emphasis of the attributes of its geographically associated origination in Silicon Valley wrapped up with its returning co-founder Steve Jobs were critical to reconstructing Apple's meaning and value: 'Steve created the only lifestyle brand in the tech industry. There are cars people are proud to have – Porsche, Ferrari, Prius – because what I drive says something about me. People feel the same way about an Apple product' (Larry Ellison,

Figure 6.10 Indirect distribution channels, Shanghai, China. Source: Author's images 2013.

Figure 6.10 (*Continued*).

CEO, Oracle, quoted on Isaacson 2011: 332). Embodying the attributes of the renewed brand, the iMac's commercial success set Apple onto a vigorous new growth trajectory through the 1990s and 2000s. Net sales grew to $42 billion in 2009 and profitability was maintained by differentiated premium pricing. Apple generated over 35% of operating profit in the PC market from only 7% of the revenue, and its share price and market capitalization recovered: 'its stock had skyrocketed to $6.56, or 33%, to close at $26.31. ... The one-day jump added $830 m to Apple's stock market capitalisation. The company was back from the edge of the grave' (Isaacson 2011: 326) (Figure 6.10).

Disruptions to regulating the accommodation between the brand's meaningful and valuable geographical associations in its Silicon Valley origination and the 'global' reach and aspirations of the actors involved in Apple have emerged recently. First, the differentiated attributes of the Apple brand have proved difficult to protect in the face of contest and rivalry by competitors. In what David Yoffie and Penelope Rossano (2012: 12) term the 'patent wars ... everyone in the industry sued everyone'. User interface and simplicity characteristics central to Apple's brand equity are being challenged by Google and Samsung, with new smartphones competing with the iPhone as well as Amazon and Samsung in the tablet market (Yoffie and Rossano 2012). Difficulties first experienced in the 1980s have replayed through the brand's history: 'As Apple found out, the "look and feel" of a computer interface design is a hard thing to protect' (Isaacson 2011: 179). Second, Apple's dramatic financial turnaround and rapid recovery has garnered increased scrutiny and questioning of its strategy in the investment community. Financial interests are rivalling the Silicon Valley, high-tech interests in the management and direction of Apple Inc. because 'as a listed company with its market capitalisation and cash issues returning it to shareholders it's becoming more Wall Street-related' (Brand Web Site Editor and Publisher, author's interview, 2013). Third, the growing international reach and market awareness of the brand, especially in emergent market contexts in Brazil, China and Indonesia where IP regulation is notoriously weaker, has raised concerns about counterfeit Apple branded products. Over 20 fake retail outlets were discovered in Kunming City, Yunnan, China, using the Apple brand and logo (BBC 2011). A whole industry supplying accessories to Apple branded products has been spawned through both legal and illegal retail channels (Figure 6.10). Apple's brand logo has permeated subcultures and been appropriated to shape other goods and services, for example in Shanghai, China (Figure 6.11).

The Apple brand's value as an intangible asset has underpinned its regulation and protection as a financial asset in the context of its growing international commercial reach. From its founding in the spatial and temporal hearth of the high-tech business in Silicon Valley in the late 1970s, intellectual property and trademarks have been utilized to safeguard Apple's technologies and differentiated attributes. The company engaged the capital markets after its founding to support initial growth and R&D investment. Apple's distinctively Californian and West Coast culture and outlook framed particular relationships with the investment community centred in Wall Street, New York. Commercial collapse and financial crisis followed its looser regulation and licensing of the brand in its unsuccessful foray into the mass market in the 1990s.

Figure 6.11 Cultural diffusion of the Apple logo, Shanghai, China. Source: Author's image 2013.

Apple's position was salvaged by the protection central to its revitalization strategy from the late 1990s in the 'Digital Hub', new mobile devices and design-led renewal of the meaning and value of the brand. Competitors tussled with Apple management over claims to ownership and proprietary of the distinctive features of the new branded products and services. Investors and shareholders haggled over the distribution and utilization of the financial outcomes of growth. Concerns have grown about appropriation of the Apple brand by external interests in weakly regulated international markets.

Summary and conclusions

Apple is a 'global brand' (Hollis 2010: 25), for some analysts, because of its commercially successful establishment of 'a consistent global brand appearance at all levels' (Lindemann 2010: 113). Origination demonstrates how the meaning and value of geographical associations in Silicon Valley provided integral attributes that enabled the actors involved in the Apple brand to construct its distinctive differentiation and international commercial expansion. Producing, circulating, consuming and regulating the 'global' in and through the Apple brand are shaped by and involve geographical associations that are rooted in the particular place of 'Silicon Valley' whether defined as a local, sub-regional or regional scale. The participant actors in the Apple brand tell 'stories of place' in its history, 'identifying with the spirit of time and place' in Silicon Valley (Beverland 2009: 57). The actors involved have originated Apple as relational and 'global' in its international reach *and* territorial and 'local', 'sub-regional' or 'regional' in its deeply rooted geographical associations in Silicon Valley, California. Apple's design, look and feel are common across countries internationally, while its ethos and identity are grounded in a particular place.

The meaning and value of Apple was initially more strongly originated in the specific national territory of America in its early expansion. Competition, internationalization and technological shifts disrupted Apple's growth path in the later 1980s and 1990s. Pressures to reduce costs underpinned the geographically uneven decentralization of production and assembly activity, relocating out of Silicon Valley initially to southern states of America and Europe. Sales decline precipitated financial losses that forced Steve Jobs' exit in the mid-1980s, heralding a period in which the differentiated meaning and value of the Apple brand was weakened and its commercial fortunes floundered. The renewal of the origination of the brand's revolutionary and visionary ethos and practice began following Steve Jobs' return in the mid-1990s. Revitalization focused on producing the 'global' in Apple: reframing the PC as the Digital Hub and late mover market entry across a range of mobile devices; renewal of the culture and ethos of the brand; design-led differentiation; reorganization of its global value chain and outsourced production strategy with lead contractors in east Asia and especially China. In circulation, a more open marketing approach was adopted, launch events became the spearhead of articulating and communicating Apple's brand, the branded 'ecosystem' of a streamlined range of Apple products and services was emphasized, and the mobile devices led the renewal of the brand meaning and value through innovative marketing internationally. In consumption, differentiation-led revitalization was focused upon late mover entrance to mobile device markets with intensively advertised and branded communications, tight integration between complementary products and services across hardware and software, and the switch from indirect to direct retail through an international network of Apple branded stores. In regulation, engagement with the capital markets and protection of the financial asset of the brand and its IP have been increasingly important in fast moving, fluid and intensely competitive and internationalizing spatial and temporal market settings. As the production, circulation, consumption and regulation of meaning and value in the Apple brand has been reinvigorated, the participant actors have had to acknowledge and articulate the shifting origination of its products and services. The actors involved have begun to experiment subtly with origination reflecting the spatial organization of its different kinds of operations. The '*Designed by Apple in California. Assembled in China*' origination enables Apple to sustain the meaning and value of its branded product and service, and reap the wide profit margins from its combination of differentiated meaning and value, premium price *and* cost efficient production. Double page adverts in the UK press in 2013 recognized this origination and proclaimed:

> if you are busy making everything, How can you perfect anything? ... We spend a lot of time on a few great things. ... We're engineers and artists. Craftsmen and inventors. We

> sign our work. You may rarely look at it. But you'll always feel it. This is our signature. And it means everything. Designed by Apple in California. (Apple advert, *The Guardian*, 27 June 2013: 24–5)

This origination appears flexible too, manifest only as '*Designed by Apple in California*' on branded products offered for sale in particular market times and spaces such as China. These uneven and unequal geographical associations hold very different implications for people and places.

Apple's contemporary commercial dominance and resonance remains under pressure as the actors involved continue their attempts to create and fix meaning and value in spatial circuits. A range of related concerns threaten disruption: intensified competition, innovation and the brand's 'jackpot' business model (Froud *et al.* 2012: 20); criticism of Apple's outsourced production model in the context of the American Federal Government's economic recovery plans following the global financial crisis; dissent from those interpreting the Apple brand and its practices as symbols of unsustainable global capitalism; maintaining its loyal and longstanding brand community amidst its international expansion; refining the brand's origination in different spatial and temporal market contexts; and demands from owners and investors to distribute excess profits, continuing regulatory battles over IPR and digital content ownership and licensing, and growing counterfeits in international markets.

Chapter Seven
Territorial Development

Introduction

Amongst the actors involved in seeking to originate brands and branding through the construction of geographical associations in branded goods and services are those concerned with the development of particular territories at a range of different scales from the supranational to the community. While their aims and aspirations in addressing the question of 'what kind of territorial development and for whom?' may differ (Pike *et al.* 2006), such individuals and institutions are implicated in differing ways and to varying degrees in the spatial circuits of meaning and value in branded commodities. Commercially successful brands strongly geographically associated with particular places are often identified as assets, lauded and promoted by actors as envoys and markers of capability and reputation. Conversely, less successful, weaker or failing brands connected to a place can be denoted as liabilities, ignored and written out of the story told of what the place is about and what the actors in it are able to do.

Territorial development actors face similar challenges to other producers, circulators, consumers and regulators in trying to cohere and stabilize meaning and value in specific brands and their branding in particular spatial and temporal market settings. They have to confront the added complications and difficulties of attempting to work with other brand and branding actors to extract benefits for a territorially defined area from scalar and relational spatial circuits. Contributions from brands to territorial development comprise harder, more tangible things including output, multipliers, investments, innovations and jobs as well as softer,

Origination: The Geographies of Brands and Branding, First Edition. Andy Pike.
© 2015 John Wiley & Sons, Ltd. Published 2015 by John Wiley & Sons, Ltd.

less tangible aspects such as capability, renown and reputation. As the analyses of Newcastle Brown Ale, Burberry and Apple demonstrated, the originations of brands in places yield benefits for particular territories in specific time periods. But the disruptive logics of accumulation, competition, differentiation and innovation mean such geographical associations are unstable and sometimes only temporary accomplishments. The geographies of brands and branding change, generating implications for specific territories connected to the brands whether they are integral to their particular development strategies or not.

This chapter focuses upon the implications of origination for territorial development. First, how the origination of brands and branding shapes the different kinds of geographical associations involved and patterns where specific economic activities take place is explained; influencing the locations of investments, jobs, supply chains, distribution channels, retail networks, trademark protections and so on in the economic landscape. How the geographical associations used by actors to originate branded goods and services commodities have enabled their role as envoys of place and markers of capability and reputation is discussed, connecting to the emergence of the brands and branding of spaces and places. The ways in which the moments of production, circulation, consumption and regulation of brands and branding in spatial circuits of meaning and value relate to territorial development are then outlined. Connections are made to the 'smile' curve of value creation and its locational implications for different types of economic activity (Chapter 3) and the potential of regulatory devices such as geographical indications in attaching brands to place for particular kinds of territorial development. Second, the chapter examines the potential of origination in territorial development. It analyses how strong originations in brands and branding utilizing geographical associations to particular places have been deployed as indigenous and endogenous assets in attempts to grow, anchor and embed economic activities for development locally and regionally. Third, the chapter explains the limitations of origination in territorial development. It demonstrates how and when strong geographical associations have become liabilities, generating lock-ins within development paths and inhibiting adaptive capacity regionally and locally.

Using origination to understand the ways actors involved in brands and branding are shaping – and in turn being shaped by – their geographical associations provides a means to lift the 'mystical veils' (Greenberg 2008: 31) they weave around branded goods and services commodities and to consider what they mean for territorial development. Origination advances beyond the development assumptions framed by 'Country of Origin', whereby the agency of actors and institutions was focused on 'national' development projects and 'national' brand champions (van Ham 2008). Origination demonstrates how relational and territorial geographical associations complicate attempts by actors to create and stabilize meaning and value in spatial and temporal market settings. Social and spatial inequalities can be (re)produced through the geographical associations constructed in brands and branding by the participant producers, circulators, consumers and regulators.

Origination in territorial development

Origination means how and why actors try to construct geographical associations in branded commodities and their branding in efforts to create and fix meaning and value for specific goods and services in certain spatial and temporal market settings. Sometimes aligned with and reinforcing the interests and work of other brand and branding actors, those involved in territorial development attempt to shape how branded goods and services commodities are originated in order to yield beneficial contributions to particular territories at different scales. Such activities involve a range of institutions with varying differing authority, power and resources working across and within different geographical spaces, ranging from inter-governmental bodies at the international level such as the OECD to organizations at the community level such as voluntary associations (Table 7.1).

Recognized as cues central to 'Country of Origin' and 'Country of Origin of Brand' effects (Chapter 3), the geographical associations integral to the origination of branded goods and services commodities have become the focus of the agency of actors in pursuit of territorial development of various kinds. Yet they too are compelled and constrained by the logics of accumulation, competition, differentiation and innovation that continually disrupt their spatial and temporal fixes of geographical associations. Changes in the economic fortunes of specific goods and services brands can affect the prospects of particular territorial economies. The implications are amplified when the brands constitute sizeable and significant specializations within regional and local economic structures and/or are closely intertwined with wider perceptions of the capability and reputation of the place. Territorial development

Table 7.1 Scales of geographical associations in brands and branding and territorial development institutions

Scale	*Examples*
Supra-national	OECD, WTO, European Commission, Economic Commission for Latin America and the Caribbean
National	Ministry of National Integration (Brazil), Ministry of International Trade and Industry (Japan)
Sub-national administrative	Bavarian Ministry of Econmic Affairs, Infrastructure, Transport and Technology, Ministry of Economic Development and Innovation (California)
'National'	Government of Catalonia, Scottish Government
Pan-regional	The Baltic Sea Region, The Northern Way (England)
Regional	Rhône-Alpes Regional Assembly, Silesian Voivodship
Sub-regional or local	Bay Area Houston Economic Partnership, Downtown Center Business Improvement District (Los Angeles)
Urban	Municipality of Milan, Council of Paris
Neighbourhood	Neighbourhood Economic Development (Portland City Council), Canadian Community Economic Development Network
Street	Saville Row Bespoke Association, Madison Avenue Business Association

institutions have to confront tensions and accommodations in trying to ground or territorialize geographical associations that are meaningful and valuable to brands and branding while they are always in flux because:

> the bounded jurisdictional spaces of governance in which many regulatory practices are established and implemented – or at least framed – by the state system, both interrupt and transform and, at the same time, are interrupted and shaped by the changing relational geographies of flows of value. (Lee 2006: 418)

The geographical associations used by the actors involved to originate brands and branding unfold in 'bounded', territorial, *and* 'unbounded', relational, space and place over time.

Different kinds, degrees and natures of geographical associations can be understood to have varying connections and implications for territorial development. Brands with strong geographical associations and a strong origination in a particular place can generate direct economic outcomes for 'development' – however defined (Pike *et al.* 2007). Such contributions can include direct, material and economic things like output, investment and employment. Indirect outcomes may occur through multipliers in purchases of goods and services and output, investment and employment amongst suppliers whether local, regional or further afield. Where ownership and control of brands with high economic values and international reputation reside, substantive power over economic resources can be centralized and concentrated in particular places. These are the sites where people make decisions about the locations of the investment projects, jobs, supply contracts for goods and services, logistics networks, retail channels intellectual property protections and so on that shape the economic landscape. Sometimes their location overlaps with key centres of government and governance in the world or global city networks (Taylor 2004), enabling dialogue and relations with key regulatory and standard setting bodies. Taking the top five in Interbrand's Best Global Brands Top 100 from 2012, the headquarters locations of the brand owners put such places on the global economic map: Coca-Cola in Atlanta, Georgia; Apple in Cupertino, California; IBM in Armonk, New York; Google in Mountain View, California; and Microsoft in Seattle, Washington. Some commentators go even further in discerning positive causal links between brands and territorial development. Steve Hilton (2003: 49) argues that the highest value brands are domiciled in the United States, Japan or EU not because brands are attracted to rich countries but because rich countries are rich because of their brands: 'Without brands, modern capitalism falls apart. No brand: no way to create mass customer loyalty; no customer loyalty: no guarantee of reliable earnings; no reliable earnings: less investment and employment; less investment and employment: less wealth created; less wealth: lower government receipts to spend on social goods.' The origination of brands and branding generates not only material and economic outcomes with implications for territorial development. Symbolic, discursive and visual forms of geographical association are used to communicate and convey meanings and values about places, their current development status and aspirations (Olins 2003). Such geographical associations are also utilized by actors to signal the future path of development a place may be seeking to follow. Brands can provide symbols, images and narratives of prosperity, success and vibrancy for actors with interests in territorial development.

When commercially successful and/or internationally recognizable brands are originated in a place then territorial development actors typically articulate and project this close and historical geographical association in positive ways as part of efforts to promote further development. 'The Home of…' format is used to suggest that a territory has an illustrious economic history by intertwining it with the locational backgrounds, stories and successes of a particular brand (Holt 2004). Newcastle Brown Ale's owners and managers were able to state and sustain this story for the brand for a period prior to its relocation from the city of Newcastle upon Tyne (Chapter 4). While actors involved in Burberry continue to originate the brand in the 'Britishness' of its national historical evolution (Chapter 5) and Apple's managers seek to project its Silicon Valley history onto the global stage (Chapter 6). When the brand is not originated regionally or locally but has been attracted and embedded from outside the territory, this can be captured and communicated as a positive endorsement of what the territory has to offer in terms of the critical factors influencing location decisions – including land and premises, skilled labour, infrastructure, tax arrangements, grants and innovation support (Markusen 2007). This strategy is used especially when territories have landed international brands that, at least in principle, have an extensive map of global locational possibilities to consider in siting their operations. Many agencies with responsibility for attracting inward investment list the existing companies attracted, typically with images and even quotes from the managers and workers at high profile brands that have located operations within their territories. This successful track record is then articulated in a bid to attract further investors to commit and associate themselves with the existing firms and brands already located in their territory. Actors in Somerset in England, for example, use this approach in highlighting the reputable brands geographically associated with the county in their promotional and economic development strategy (Figure 7.1).

The geographical associations – material, discursive, symbolic and visual – that connect branded goods and services to places relate to the voluminous and growing literature on the creation of space and place brands and their branding. Extending from the commercial world of goods and services brands and branding, such space and place brands are constructed by actors in international competition between territories to attract, embed and retain developmental resources and opportunities across an expanding field of activities. Such brands now encompass competition for new and existing businesses, investments, jobs, occupations, residents, skilled labour, spectacle events, students and visitors (see, for example, Greenberg 2008; Hollands and Chatterton 2003; Ashworth and Kavaratzis 2010; Lewis 2007; Pike 2011b). In a competitive context where territories are increasingly establishing brands and proactively engaging in branding activities, places that try to opt out risk becoming invisible, ignored and left behind. To paraphrase Nicola Bellini (2011), all places have a reputation; whether they choose to manage it or not is the key question. Space and place brands and branding now occupy a significant part in the identity and image management elements of territorial development and play an integral role in

Figure 7.1 'Somerset – The natural choice for business'. Source: Into Somerset.

transformative projects focused upon 'rebranding' territories to stimulate new developments (Halkier and Therkelsen 2011; Pike 2011b).

Space and place branding remains though an emergent and still growing field beset by fundamental and unresolved issues. These concerns include: definition, conceptualization and theorization (Go and Govers 2010); how and why place brands are constructed and by whom (Richardson 2012); the (mis)translation of brands and branding from goods and services to more complex spaces and places (Pike 2011d); tensions between the search for distinctiveness and tendencies to homogenous approaches (Turok 2009); how space and place brands become embedded and sustained over time (Richardson 2012); identification and assessment of their contributions and effects (Richardson 2012); how the brands interrelate with the wider notion of reputation (Bell 2013); and the politics of space and place brand ownership and representation (Aronczyk 2013; Greenberg 2010; Julier 2005).

In examining how the production, circulation, consumption and regulation of brands and branding in spatial circuits of meaning and value have territorial development implications, it is instructive to revisit the 'smile' curve of value creation (Figure 3.4) and examine its implications for location and development prospects. Ram Mudambi (2008: 706) demonstrates that 'The geographic realities associated with the smile of value creation are that the activities at the ends of the overall value constellation are largely located in advanced market economies, while those in the middle of the value chain are moving (or have moved) to emerging market economies' (Figure 7.2). Actors involved in different activities within the global value chain face different contexts and incentives categorized as 'catch-up', 'spillover' and 'industry

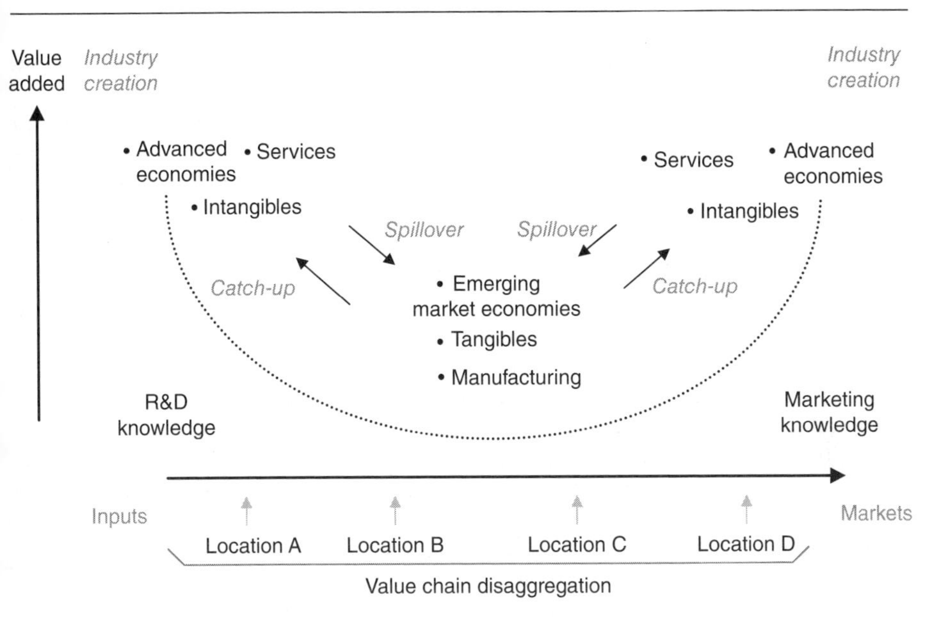

Figure 7.2 Value chain creation and location. Source: Adapted from Mudambi (2008: 709).

creation' (Mudambi 2008: 708). Catch-up refers to actors situated in the middle of the global value chain and based in emerging market economies – such as Brazil, China, India and Mexico – attempting to upgrade their positions through developing the resources and capabilities to undertake and control higher value-added activities. Such endeavour includes the development of own brands and marketing expertise to enhance their presence and value capture in market settings downstream as well as the bolstering of their R&D and innovation capabilities upstream. R&D, marketing and branding functions have been located in advanced markets as listening posts to inform their development and as mechanisms to increase their ability to pick up and absorb new ideas, fashions and trends from particular market times and spaces settings.

Spillover occurs as actors from advanced market economies controlling the upstream and downstream ends of the global value chain reorganize and relocate even higher-value added activities to emerging market economies to improve their overall cost efficiency. Under increased competitive pressure, especially from the new market entrants undertaking catch-up activities, existing actors have tried to:

> increase the efficiency and effectiveness of the high value-added activities that they control. Modularization enables these firms to strip out standardized activities from both the upstream R&D and downstream marketing activities that can then be relocated to emerging market economies. (Mudambi 2008: 709)

Innovation at each of the high value ends underpins new industry creation emerging from 'basic and applied R&D at the upstream end (e.g. biotech, nanotech) and through marketing and distribution innovations at the downstream end (e.g. e-tailing, online auctions)' (Mudambi 2008: 709).

Relating changes in the smile curve to the spatial circuit of meaning and value, in the production moment, assembly and manufacturing activities have been and continue to be relocated from advanced to emerging economies to different degrees and in different ways in different economic activities (Dicken 2011). Places losing the activities of assembling and producing branded goods or their constituent components face territorial adaptation challenges in coping with the lost output, investment and employment associated with deindustrialization (Pike 2009c). The places gaining and/or growing assembly and manufacturing activities confront issues of how to support the emergence and often rapid growth of new economic activities, ensuring the capital, infrastructure, skills and innovation support is in place to make lasting developmental gains and provide the basis for future upgrading (Yeung 2009). In South Korea, for example, an increasing emphasis is being placed upon design – merging the ideas of 'design' and 'economics' (Figure 7.3) as a means of constructing meaning and value for goods and services brands within upgrading strategies.

In the context of the smile curve, circulation and consumption have grown in importance. Branding and marketing have been emphasized for upgrading economic activities and adding value and meaning to goods and services commodities. At the

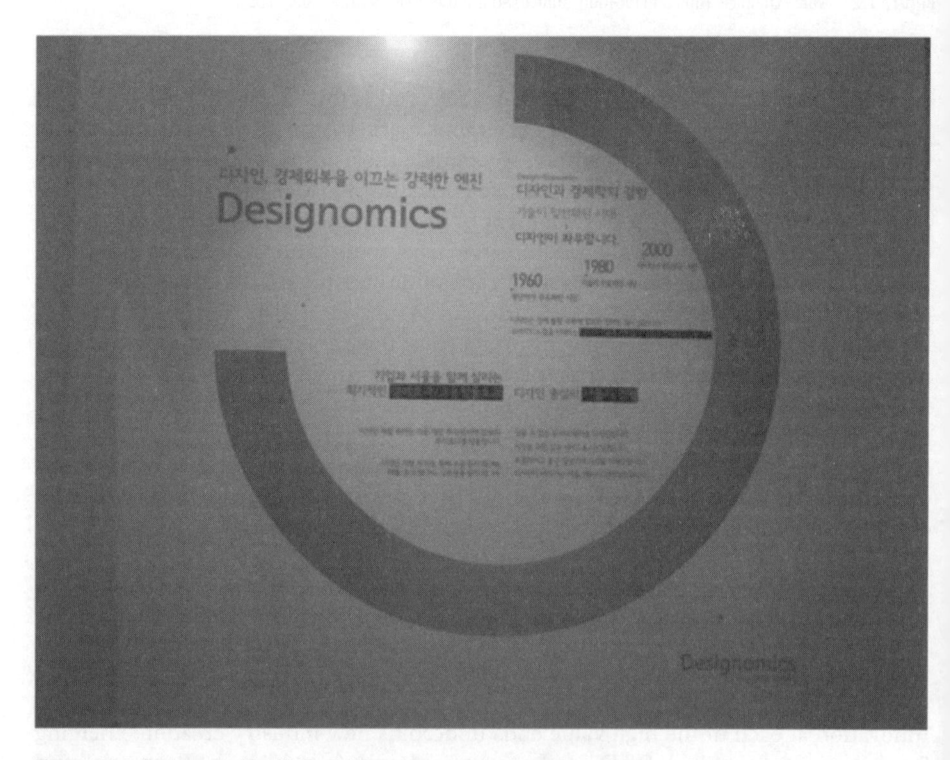

Figure 7.3 'Designomics', Seoul, South Korea. Source: Image by Mário Vale 2011.

top end of the spatial hierarchy, the globally influential centres of advertising, branding and media in London, New York and Tokyo have benefitted from the growing importance of brands and the world of branding. Enduring positive outcomes have accrued and been reinforced in such cities through continued flows of investment and contracts, ongoing creation of high-value and highly remunerated occupations, and extended global influence (Taylor 2004). The brands of such advertising and media giants are embellished by their close geographical associations in places such as Soho in London, Madison Avenue in New York, and Shibuya and Shiodome in Tokyo (Faulconbridge *et al.* 2010, Grabher 2001).

In the emerging economies, actors are engaged in trying to upgrade through the development of R&D and marketing and branding capabilities that 'generate negative cashflow in the short run as resources are withdrawn from low margin contract manufacturing and assembly or standardized service delivery and transferred to R&D and marketing where the firm has little experience' (Mudambi 2008: 708). But initial involvement in the lower value-added activities resulting from outsourcing from advanced economies as well as indigenous development are vital early steps, as well as investments in developing capability and reputation as a platform to receive further spillover functions from advanced economies and to enable efforts at further upgrading (Barrientos *et al.* 2011).

The regulation moment in spatial circuits of meaning and value in brands and branding is particularly important in protecting, valorizing and sustaining the geographical associations of specific branded goods and services in certain market times and spaces. State, para-state and independent institutions act as jurisdictional territorial entities across a range of geographical scales to authorize, control and protect the legal status, ownership and geographical associations of brands through each moment from production through circulation to consumption. Brand establishment and protection are covered by different national structures of trademark law that afford unrestricted monopoly rights and, unlike patent and copyright, render brands excludable for unlimited time periods. Words, colours and signs are difficult to protect in trademarks if liable to general trade use (Lury 2004). Geographical associations can provide the distinctive differentiation central to trademark strength, for example in ingredients, packaging or names. Where origin of production is central to brand equity and value, strong protection is required to prevent imitation by co-location. Significantly in relation to the tensions and accommodations between territorial and relational geographical associations in origination, current trademark definitions contain an implicitly unbounded sense of geographical attachment: 'the distinctiveness of a logo is increasingly being judged in terms of linkages and associations, and not ... in relation to a fixed origin' (Lury 2004: 14).

Other regulatory devices provide more explicit ways in which the geographical associations of branded goods and services can be defined, protected and connected to territorial development. Geographical Indications (GIs), in particular, are marks that 'can be seen as attempts to tie particular qualities inherent in the product to particular qualities inherent in the context of production' (Parrott *et al.* 2002: 246). GIs are used to establish, reinforce and strengthen the geographical associations of brands and branding in place for territorial development by potentially 're-linking production to the social, cultural and environmental aspects of particular places ... and opening the possibility of increased responsibility to place' (Barham 2003: 129). GIs contrast the securitization of geographical associations in brands through trademarks which convert attributes of place and local knowledges into property, rendering such associations private, tradeable and vulnerable to de-territorialization (Morgan *et al.* 2006). For Elizabeth Barham (2003: 129), 'as a form of collective property anchored to specific places, GIs challenge ... globalization ... [and] a frictionless economy where neither space nor time impedes the free flow of goods, labour and capital'. The place attachments and regulation of place in GIs means that:

> Unlike trademarks, which are privately owned intellectual property rights that can be bought and sold, GIs are a spatially specific public good, in the sense that they protect the geographical name of a product from a given region. The collective legal status of GIs helps small producers ... to gain a form of protection and promotion for their products that is embedded in, and tied to, their region, an asset that, unlike conventional trademarks, cannot be delocalized. (Morgan *et al.* 2006: 186)

The EU's Protected Geographical Indication (PGI) designation for agro-food brands – such as Aberdeen Angus beef and Parma ham – is one regulatory scheme that has explicitly linked the regulation of branded product provenance and quality to local development (Parrott *et al.* 2002). Such GI schemes and their territorial development potential are receiving attention internationally, including Antigua coffee in Guatemala, Darjeeling tea in India and Etivaz cheese in Switzerland.

The same regulatory devices capable of generating positive contributions to territorial development can be negative in certain settings. Territorially bounded forms of the regulation of geographical associations in brands create and legally protect spatial monopolies and rents, barriers to market entry and supply controls: 'differentiating the isotropic space of classical rent theory. By restricting action and interaction, they give advantages to some areas and exclude other areas' (Moran 1993: 695). Regulation is socially and spatially uneven in its ability to afford protection for some brands against appropriation in non-local markets. However, it can also reduce brands to registered names and risk the erosion of their geographical associations with place and quality. Despite the WTO's ruling that GIs were compatible with its rules and could coexist with prior trademarks, GIs and other place regulation in brands (e.g. 'Country of Origin Labelling') remain political battlegrounds between US-led interpretations of protectionism and EU-led support for localized protection against inferior imitation (Hayes *et al.* 2004; Morgan *et al.* 2006; Parrott *et al.* 2002).

The potential of origination in territorial development

Where actors involved in brands and branding make strong geographical associations to particular places as part of their attempts to originate goods and services commodities and construct differentiated meaning and value, then they provide a means to attract, anchor and embed economic activities in place and contribute to determinations of territorial 'development' (Pike *et al.* 2011). Brands with the strong and enduring geographical associations between branded commodities and places support the creation of what Colin Crouch (2007: 211) calls 'collective competition goods' by local and/or regional institutions, including regional and local innovation systems, skills development infrastructures, supply chains and bespoke sites and premises. These collective and territorially embedded assets can be developed in creating the added value and meaning used 'to sustain the cultural-*cum*-economic virtues of such places and to safeguard their products and reputations from the negative influence of cheap imitations' (Scott 1998: 109). Brands and their branding strongly embedded and originated in place can be deployed as indigenous assets underpinning (endogenous) growth from within localities and regions as well as (exogenous) export growth and the attraction and embedding of external resources such as investment and people. In originations where the actors mobilize strong geographical associations to particular places this is used to underpin strategies for 'territorial (re)valorization' (Ray 1998) whereby 'well-known regional products ... act as a "flagship" for a region, creating ... other synergies in terms of tourism and regional development ... and ... a broader valorization of place through asserting traditional cultural identities' (Parrott *et al.* 2002: 257).

Origination of branded goods and services by actors based upon strong geographical associations in place provide a range of potentially positive outcomes for development of the territories within which they are based and/or associated with. Producers and deliverers of services generate output for sale within and beyond their territories. This involves assembling the means of production with capital, labour and land. Economic activity generates both direct and indirect demand for further goods and services that is territorialized to varying degrees at different scales and within different circuits and networks. In some cases, the distinctive differentiation underpinned by origination supports increased and locally retained sales and income, generating virtuous and cumulative growth (Ilbery and Kneafsey 1999; Tregear 2003). A demanding and sophisticated 'local consumption constituency' (Molotch 2002: 668) provides the critical mass for such originated brands to grow, supporting specialization, geographically proximate producer–consumer interaction and exports. Economic activities with strong originations in place may attract investment, building the capital stock and enabling productivity growth within regional and local economies. Employment can be created across a range of occupations and levels, sometimes in highly skilled, specialized, even artisanal and craft work (Tregear 2003). When owned by regional and/or local interests, economic activities can be more embedded and committed to their domicile regional and local economy and less likely to relocate (Ilbery and Kneafsey 1999). Economic entities that have deeper and stronger geographical associations in a particular place and rely upon its attributes to originate the meaning and value in their branded goods and services may be more able or disposed towards pursuing more sustainable business practices in their employment and sourcing. Businesses with strong geographical associations originated in place could even stimulate brand extensions and spin-offs through providing assets for new business activities.

Connected to the harder, more tangible developmental outcomes for territories are the softer, less tangible potential benefits of branded goods and services originated with strong geographical associations to particular places. Michael Beverland (2009: 149) argues for positive linkage between brands and places, reinforcing the authenticity of each:

> The brand benefits because it gains from the unique characteristics of the region, tourists travelling through, and regional publicity. Likewise, the region and local communities benefit from active regional players. As a result, these brands further embed themselves in the local landscape, all of which enhances their authenticity.

Brands with strong geographical associations in place serve as envoys and promoters of place in international territorial competition (van Ham 2001). Actors involved can find that 'the promotion of quality brands can segue into the promotion of a specific territory, its products and its producers' (Morgan *et al.* 2006: 102). Attributes of particular places embodied to varying degrees become mobile through strongly geographically associated brands: 'The local has more consequence because it can so vastly move out through the goods to so many other places' (Molotch 2002: 686). Such brands provide the focus and vehicle for civic pride and vibrancy within localities and regions (Ilbery and Kneafsey 1999; Tregear 2003).

Strong geographical associations in the origination of specific brands and branding serve to anchor economic activities in places by rendering mobility costly and inhibiting spatial switching and the substitution of geographical associations for those with potentially different and/or even less value or meaning. Strong geographical associations are especially evident when the brand's integral elements are spatially fixed, or not cost-effectively or perfectly substitutable elsewhere. Such geographical associations are often evident in agro-food brands: 'because the various mixtures between the organic and inorganic are hard to detach from space and place' (Morgan *et al.* 2006: 10). When a brand's integral characteristics – design, provenance, quality – are rooted in place then such attributes may be non-replicable (Parrott *et al.* 2002). A brand then cannot be made elsewhere beyond a specific spatial threshold without compromising its distinctive meaning and value. In the case of French wine, for example: 'the *context* of production – culture, tradition, production process, terrain, climate, local knowledge system ... "terroir" ... strongly shapes the quality of the product itself' (Parrott *et al.* 2002: 248; emphasis in original). Material and symbolic geographical associations to particular places form integral elements of such wine brands: 'their very existence and name cannot be separated from the production that is practised within their territory ... place names and production are ... inseparable' (Moran 1993: 698). Strong geographical associations are evident in other manufactures without such intrinsic or inherent technical ties to specific sites of production. Acting as a territorial anchor, for example, the origination of branded Swatch watches shapes its economic geographies and territorial development potential: 'This guarantee requires – or at least implies – the use of Swiss labour in the manufacture of Swatch products. It thus limits the extent to which the Swatch company is able to move production of Swatch-branded products to take advantage of lower labour costs outside Swiss national territory. ... It suggests that Swatch is best seen as a national, territorial brand' (Lury 2004: 54).

An enduring and longstanding origination related to territorial development is based on the strong and embedded geographical associations of Harris Tweed with the Isle of Harris in the Outer Hebrides, western Scotland, 'from which the Tweed takes its name' (*Daily News*, 1 August 1906 cited in Hunter 2001: 12). Amidst cyclical growth and lower output since its peak in the 1960s, Harris Tweed has been 'a vital and integral part of Island life for over a century' (Hunter 2001: 14). The Outer Hebrides and Harris are remote rural island economies facing several challenges: dependence upon relatively few economic sectors with several affected by seasonality and/or vulnerability to economic cycles and shocks; out-migration of the economically active population; and dependence upon the public sector for employment, transfer payments such as social security and state pensions, and capital grants (Comhairle nan Eilean Siar 2003). In this context, Harris Tweed has been a vital source of incomes and livelihoods for the local population.

Regulation has been critical to the survival and protection of the meaning and value of the brand. Critical has been avoiding its dilution into a generic term with little or no connection to the place in its brand name. This situation befell Cheddar cheese, severing its historical links to Cheddar, Somerset, England. The main threats to Harris Tweed have been twofold. First, throughout its history it has been subject to frequent episodes of 'passing off, or imitation of the genuine article by unscrupulous manufacturers outwith the Hebrides' (Hunter 2001: 13). Second, the brand has been damaged by 'those who allowed inferior cloth to be sold as "Harris Tweed", in order to meet demand for increased production' (Hunter 2001: 14). Together, the threats to disrupt the meaning and value of the brand in its particular spatial and temporal market contexts would have detrimental impacts upon its contributions to territorial development in Harris. If either fakes or inferior product 'had been allowed to flood the textile market, the genuine article would have suffered and a vital source of income for the people of the Outer Hebrides would have been lost' (Hunter 2001: 13–14). Origination has been integral to the regulated definition of Harris Tweed. It secures and emphasizes its attachment and tie to the Isle of Harris by stating 'it should be handwoven by the weavers at their own homes from 100% new wool and spun, dyed and finished in the islands' (Hunter 2001: 344).

Institutional innovations have underpinned regulation and territorial development efforts, initially with the Harris Tweed Association (HTA) in 1909 and the statutory Harris Tweed Authority in 1993. The HTA was established by an Act of Parliament with responsibility for 'promoting and maintaining the authenticity, standard and reputation of Harris Tweed' (Hunter 2001: 14). For the actors involved:

> The intention was that the Act would protect the (intellectual) property in the name 'Harris Tweed', a local asset ... Harris Tweed was a local resource, there was vested property in it for the purposes of the legislation, it had a reputation, it had a store of goodwill which was the collective property of the community in the Western Isles. (Harris Tweed Association Minutes, 30 March 1990, cited in Hunter 2001: 351–2)

The Harris Tweed Authority was vital in securing, protecting and defending the Orb trademark for the brand to 'guarantee any customer that he was buying genuine, good quality, handspun Harris Tweed' (Hunter 2001: 68) (Figure 7.4). Recent circulation efforts have sought to connect Harris Tweed to mainstream fashion and other high-profile clothing brands such as 'Top Man and preppy American retailers like J. Crew, and can be found covering headphones, holdalls, North Face jackets, and Dr. Martens

Figure 7.4 Harris Tweed trademark and product label. Source: Personal Communication with The Harris Tweed Authority.

Figure 7.4 (*Continued*).

and Converse boots' (Carrell 2012: 13). Links to broader promotion of the Isle of Harris and Outer Hebrides connect to the famous and world renowned brand, cemented by opportunities to buy authentic Harris Tweed from outlets on the island (Figure 7.5).

Attempting to mimic the local and regional benefits of such longstanding originations of branded goods and services, actors elsewhere have sought to originate and bring into being new brands and branding as parts of territorial development strategies. In Castilla-La Mancha, one of the poorest regions of Spain with a GDP of €18 334 (PPS) per capita (82% of EU27) in 2005 (Eurostat 2008), the regional government has sought to stimulate economic activities and raise the profile of the territory by establishing a quality brand of the locally grown spice saffron as a luxury and premium-priced export.

Figure 7.5 Sold in Harris, retail outlet, Tarbet, Isle of Harris. Source: Author's image 2012.

Figure 7.6 Saffron label, Castilla la Mancha. Source: Azafran de la Mancha, http://www.doazafrandelamancha.com/2/.

The local brand is seeking to exploit growing international demand for authentic provenance and quality assured agro-food products and tap into the sales success of other gourmet ingredients from Spain such as olive oil and wine (Fuchs 2006). A new brand name and identity has been created to establish its distinctive differentiation (Figure 7.6). Local production in La Mancha had been in decline for decades because of lower-cost competition from a variety produced in Iran that hitherto provided 90% of the world saffron supply. But the market position of producers in Iran has been undermined by the geo-political context of the American trade embargo.

The La Mancha regional government established a quality control board – *Consejo Regulador de la Denominación de Origin Azafrán de la Mancha* – to regulate production quality and provenance, and certificate producers with an official seal. Local producers are being supported jointly to exhibit at international trade fairs. A saffron restaurant route has been mapped out locally to encourage visitors and stimulate business and demand for saffron among local food providers. Branding activity has sought to differentiate the local saffron brand through designer packaging in distinctive tins and glass jars with the La Mancha brand label distributed through gourmet retail outlets in New York and Paris. Production has risen from 50 kg in 1999 to more than 1016 kg in 2006. The premium price strategy has led to the almost doubling of prices from €800 to €1500 per kilogramme between 2001 and 2006. This surge in demand and value has destabilized the existing market, creating a niche segment and ensuring its insulation from competition from lower-cost spice produced in Iran selling at €400 per kilogramme. Growing production of this branded saffron now provides work and incomes for more than 1000 families in the region where mostly women work to undertake the painstaking labour of harvesting the flowers and extracting and heating the filaments to produce the spice.

The limitations of origination in territorial development

The potential of brands and branding originated with strong geographical associations by actors for territorial development has to be qualified and its limitations recognized. As with other forms of economic specialization, when the fortunes of specific local and regional economies are closely tied to those of certain brands in spatial and temporal market settings they can be vulnerable. Disruptive rationales of accumulation, competition, differentiation and innovation are transmitted by and through brands from their spatial circuits to territorial economies. As a form of specialization wedded to specific brands, territorial development may suffer as a result of a lack of diversification and dependence upon a narrow set of economic assets and resources tied up with branded goods and services. Brands can encourage over-specialization and vulnerability, especially if goods and services producers end up competing against each other in the kinds of relatively small and finite niche markets that are carved out through their differentiation efforts (Watts *et al.* 2005). When the geographical associations used by actors to originate brands are weaker and less vital parts of their meaning and value, economic activities can be relocated, severing some kinds of geographical associations but not necessarily damaging the commercial prospects of the brand in specific market times and spaces as in the cases of Newcastle Brown Ale and Burberry (Chapters 4 and 5).

Brands with strong geographical associations in place embroil their respective domicile and/or host territories in competitive relations as rival brands compete for share and dominance in certain spatial and temporal market settings. 'Winners' and 'losers' in markets contribute to the geographically uneven development inscribed in brands and branding (Chapter 2). The unequal outcomes of such competition have potential implications for the ability of brand owners and managers to sustain investment and jobs within the local and regional economies in which they are situated. Peter Van Ham (2001: 2, 6) even claims that 'brands and states often merge in the minds of the global consumer ... strong brands are important in attracting foreign direct investment, recruiting the best and the brightest and wielding political influence. ... In this crowded arena, states that lack relevant brand equity will not survive.'

Brands with strong geographical associations originated in place generate the kinds of functional, cognitive and political 'lock-ins' Gernot Grabher (1993) envisaged inhibiting local and regional economic adaptation and innovation. Brands with commercially successful histories can dominate the outlook and perspectives of actors in particular territories, limiting their ability to respond to adverse changes. As powerful and enduring symbols of meaning and value, brands can constrain adaptive capacity and the capability of actors in places to interpret, identify and generate fruitful development opportunities (Pike *et al.* 2010). Developing a reputation for being good at something can pigeonhole territories and constrain the creation of new and future development paths.

The decline of Eastman Kodak and the Kodak brand with its longstanding geographical associations in the city of Rochester, New York State, demonstrates the perils for territorial development of a strong origination between brand and place. With its headquarters in the city, Eastman Kodak was Rochester's largest employer, accounting for 60 000 jobs when the city's population peaked at 330 000 in 1950 – earning the city the nickname 'Kodak Town' (NPR Staff 2012) (Figure 7.7). The geographical associations in the brand enabled an origination for Kodak situated within the era of American post-war manufacturing expansion when the meaning and value of 'Made in the USA' was an internationally powerful component of 'Brand America' (Anholt and Hildreth 2004). Kodak became an 'American icon' (Sheyder 2012: 1) providing the photographic technology to capture the growth of mass consumerism in America in the post-war period. The brand was a household name with its brand strapline 'It's a Kodak moment'. From its position as a 'top global brand' in the 1960s and 1970s when it dominated the film and camera business (Karlgaard 2012: 1), Kodak experienced declining profit margins during the 1980s. Despite its early innovation in digital cameras in the mid-1970s, Kodak feared 'cannibalizing their core film sales' and was only spurred to introduce its first own brand digital camera in 1995 (Sheyder 2012: 1). By the mid-1990s, however, the market was already being dominated by Canon, Nikon, Sony and other Asian brands such as Samsung. Mobile phones later revolutionized the market by incorporating digital camera technology.

The managers of the Kodak brand found it difficult to adapt to the shifting spatial and temporal market context. Continued film production was a highly profitable source of cash flow, new projects were abandoned too quickly, and its investments in digital technology were unfocused (Sheyder 2012). The company prolonged sales of Kodak digital cameras at a loss in an attempt to maintain market share and brand awareness. Trading conditions moved adversely for Kodak with its market share collapsing from 27% in 1999 to 7% by 2010. In response, between 2004 and 2007 Kodak closed 13 film plants and 130 photo labs, reducing its workforce by 50 000 (Sheyder 2012). Employment in Rochester fell from 19 000 in the 1980s to under 5 000 by 2012.

Figure 7.7 Kodak headquarters, Rochester, New York State. Source: Christian Scully, 2011, Corbis.com.

For some, the socio-spatial history of the brand and its strong geographical associations in Rochester bred a 'complacency' that 'blinded the company from technological leaps elsewhere' (Sheyder 2012: 1). Rosabeth Kanter went further in seeing that 'The seeds of the problems of today go back several decades … Kodak was very Rochester-centric and never really developed a presence in centers of the world that were developing new technologies. … It's like they're living in a museum' (quoted in Sheyder 2012: 1). Richard Karlgaard (2012: 1) argued too that a kind of 'paralysis can infect cities and regions' because:

> Kodak's other structural problem is geography. When you study the history of great American companies that stumbled and failed, or only partially recovered, you see how difficult it is to overcome the mindset of your immediate surroundings. Businesses

located in places where success is the norm, and innovation is built into the ecology, have a better chance of fixing themselves. Intel almost bit the dust in the mid-1980s but came back to greater glory. Like Kodak, it faced ruinous Japanese competition. Intel didn't hesitate. It shed its memory-chip business and bet the ranch on microprocessors. That was a big bet and it was ruthless. Memory-chip factories were shuttered and people were laid off. That was, and is, easier to do in Silicon Valley, where the laid-off can more readily find new jobs, than in a small city like Rochester, whose population is now at 210,000 plus.

After 130 years, Eastman Kodak filed for bankruptcy in 2012. It secured a $950 million credit facility to continue operation and maintain employment of its 17 000 workforce as it attempted to reduce liabilities of more than $6 billion and sell some of its 1 100 digital patents (Sheyder 2012). Richard Karlgaard (2012: 1) concluded that the painful restructuring was deferred because 'It would have been infinitely harder to do in Rochester, because the impact on a small city and the multiplier effect of lost jobs, axed all at once, would have been a civic disaster. Of course, Kodak's slow bleed has turned out to be a civic disaster anyway.'

During its history, Harris Tweed demonstrates not only the potential for territorial development but also the perils resulting from its origination and strong geographical associations in place. Recurrent attempts have been made for restructuring and modernization of the industry as production became more volatile and exhibited a downward trajectory from its 1960s peak, leading to closures, redundancies and layoffs. Such changes reflected the periodic collapse of the brand's coherence and stability in certain spatial and temporal market contexts specifically in America and Asia. The crisis in the 1980s underlined the fragility of the island's economy and its dependence upon the declining fortunes of Harris Tweed.

In the 1990s, the participant actors sought an integrated strategy for the industry's development. The focus was upon 'trying to re-position Harris Tweed as a niche market' (Spokesperson for Highlands and Islands Enterprise, quoted in *West Highland Free Press*, 25 September 1992 cited in Hunter 2001: 341) rather than increasing manufacturing capacity and reducing prices in a slack market, especially for exports. A 5-year revitalization plan supported by national and EU funding addressed the collapse in especially its key export markets in America and Canada. The plan involved the key actors: the Harris Tweed Authority; the Weavers' Union; the producers (sheep rearers, weavers and millers); *Comhairle nan Eilean* (Western Isles Council); Highlands and Islands Enterprise; Western Isles Enterprise; and Lews Castle College. Key elements included: addressing changing market demands for wider, softer and lighter weight cloth; amalgamating and consolidating existing manufacturers; upgrading weaving equipment; replacing the ageing workforce with younger and skilled crafts people; securing access to finance; modernizing marketing and promotion; and protecting the trademark in the European single market (Hunter 2001). While 'the name "Harris Tweed" is still known and respected across the world' (Hunter 2001: 355), ongoing pressures from accumulation, competition, differentiation, innovation and changing market fashions and tastes continue to disrupt the brand's development trajectory. Current concerns include the increased volatility and uncertainty in the textile market, trade disputes blighting continuing access to the American market, financial crises in Asia as well as internal dissension and disagreement about future development strategy amongst the actors in the industry on the island itself (Hunter 2001).

Summary and conclusions

This chapter examined the implications of origination for territorial development, connecting to the central aim of the book in addressing what origination and geographical association means, how and where it works, who does it and what it means for people and places. The ways in which the different kinds, degrees and natures of geographical associations in origination shape economic geographies and their implications for territorial development were explained. Material, symbolic, discursive and visual geographical associations have been utilized by actors in originating brands and branding as part of their efforts to construct and cohere meaning and value. Actors with interests in territorial development have tried to connect and influence such geographical associations to benefit specific development strategies for particular territories. Material outcomes such as output, investment and jobs have been sought alongside symbolic, discursive and visual signs and markers of capability, competence and reputation in the international competition amongst increasingly branded territories to attract, embed and retain resources for development. Transitions in the activities and geographical distributions of economic activities within global value chains are reshaping economic geographies and territorial development. Regulatory devices have been used as part of efforts to fix economic activities in place through protection of the geographical associations and origination of brands.

Where geographical associations are strong and commercially meaningful and valuable in certain spatial and temporal market settings, origination has demonstrated potential and has been utilized to make positive contributions to territorial development. Brands with strong geographical associations in place have constituted central assets in indigenous, endogenous and exogenous growth models and strategies. Acting as a focus, magnet and vehicle for investment, job creation and training, brands have been integral to value generation and capture within particular regional and local economies. Harris Tweed sustains livelihoods and incomes in a relatively remote and fragile rural island economy, linking it to spatial circuits internationally. Castilla-La Mancha Saffron is supporting efforts to connect a regional economy to a global value chain in the interests of territorial development, job creation and livelihood support.

Yet origination through geographical associations in brands and their branding has limitations that inhibit development within particular territories under certain conditions. Regional and local economies have become overly specialized and dependent upon the fortunes of specific brands in their specific market times and spaces. Vulnerability has been heightened by tying the territorial development fortunes of places to the vagaries of brands buffeted by disruptive logics of accumulation, competition, differentiation and innovation. Enduring and longstanding geographical associations have generated lock-ins inhibiting the potential for adaptation and the creation of new growth paths. Kodak's brand managers struggled to adapt to transformations in its spatial and temporal market context, constrained by its historical embeddedness in Rochester, New York State. Harris Tweed continues to face on-going restructuring challenges to protect, sustain and revitalize its commercial prospects in a shifting international scene while attempting to secure development locally.

The sophistication of branding practices and evolving spatial organization of economic activities mean that depending on specific connotations in particular geographical markets can be risky and ephemeral for territorial development. Origination demonstrates how brand owners now have greater albeit highly uneven potential and capacity to play up or hide origin cues, selectively constructing and representing geographical associations through branding. Brands and branding can be constructed and made to look as if they come from and connote particular places by the participant actors even when they might have quite different underlying material geographies. The geographical associations of brands and branding, then, may ultimately provide only a fragile asset capable of mobilization for territorial development and perhaps an unsustainable bulwark against the disembedding and de-territorializing forces of international economic integration. Whether weak or strong, geographical associations can be severed and economic activities relocated. As marketing constructs brands may be risky and potentially volatile foundations upon which to base territorial development strategies, susceptible to the vagaries of consumer preferences and market shifts, acquisitions and relocations – as demonstrated in the originations of Newcastle Brown Ale, Burberry and Apple and the fates of closed operations in Newcastle, Gateshead, Treorchy, Rotherham, Fremont and elsewhere. While the geographical associations of brands and branding may be inescapable, the agency of actors remains critical in selectively constructing and articulating their meaning and value in the context of specific development projects for particular territories. Geographical associations and origination provide a framework to interrogate and explain how such interests are involved and what it means for people and places.

Chapter Eight
Conclusions

Introduction

From the enduring reputation and renown of the pioneering engineering history of north east England and its contemporary resonance, where goods and services commodities are from and are associated with – and where they are *perceived* to be from and associated with – has been the central concern of *Origination*. Engaging the history and dramatic and pervasive rise of the brands and branding of goods and services commodities, *Origination* has addressed the relative dearth of attention to their interrelations with spaces and places. Although a literature is emerging, the geographies of brands and branding have been lacking conceptualization and theorization, analytical and methodological approaches, and a stock of empirical research. The need for critical enquiry has become acute. Internationalization, even globalization, has questioned notions of 'Country of Origin' in articulating the meaning and value of goods and services commodities (Phau and Prendergast 1999). 'Origin identifiers' (Papadopoulos 1993: 10) are increasingly used in marketing activities. Interest in commodity origins, provenance and transparency has grown substantially (Beverland 2009). Actors involved in brands and branding have become more sophisticated in their thinking, strategies, frameworks, techniques and practices.

Demonstrating the *inescapable* geographies of brands and branding, *Origination* develops understanding and explanation of what such geographical associations are, how and where they work, who creates and articulates them and what they mean for people and places. Origination was defined as the attempts by actors interrelated in spatial circuits – producers, circulators, consumers and regulators – to construct

Origination: The Geographies of Brands and Branding, First Edition. Andy Pike.
© 2015 John Wiley & Sons, Ltd. Published 2015 by John Wiley & Sons, Ltd.

meaningful and valuable geographical associations – material, discursive, symbolic, visual – for branded goods and services commodities. Geographical associations are utilized as part of the efforts of actors to create, cohere and stabilize meaning and value in specific brands and their branding in particular spatial and temporal market contexts. In contemporary branded 'cognitive-cultural capitalism' (Scott 2007: 1466), origination reconnected with the ideas of Karl Marx (1976) and David Harvey (1990) to counter the 'de-fetishization' critique. Origination provides a new way of lifting the 'mystical veils' (Greenberg 2008: 31) woven by the increasingly sophisticated work of brand and branding actors to better understand the economic, social, political, cultural and ecological conditions in which and where they are organized. Second, *Origination* furthers geographical theory and encourages research in other social science disciplines with an interest in the spatial dimensions of brands and branding. It illustrates the importance and worth of connection and dialogue at the intersection of political *and* cultural economy approaches to interpreting and explaining spatial circuits of meaning and value. *Origination* specifies and elucidates the roles of actors with particular interests related within spatial circuits and animated through rationales of accumulation, competition, differentiation and innovation. Such actors include not only the producers and consumers, but the circulators and regulators too. Origination explains how actors try to construct geographical associations to establish and cohere meaning and value in and through branded goods and services commodities in market settings in space and time. Buffeted by ongoing disruptive logics, such fixes are often only temporary accomplishments and require constant attention and effort to sustain and develop. Origination is variegated by the actors involved in shaping and being shaped by the different kinds, degrees and natures of geographical associations entwined in branded commodities shifting across and between spatial scales *and* within relational circuits and networks. This final chapter concludes *Origination*. First, the wider arguments and contributions of origination to the political and cultural economy of the geographies of brands and branding are articulated. Second, reflections on the politics of origination and their territorial development implications are discussed.

Origination in the political and cultural economy of the geographies of brands and branding

Origination provides conceptual, theoretical and analytical advances to understanding and interpreting the geographies of commodity brands and branding. Rather than just a descriptive metaphor, origination explains where goods and services are from and are associated with. It focuses upon how actors in spatial circuits of production, circulation, consumption and regulation attempt to construct geographical associations to particular spaces and places. Such efforts seek to create, cohere and stabilize meaning and value in brands and branding in particular spatial and temporal market settings. Origination makes several wider conceptual and theoretical contributions. First, geographical associations are defined and conceptualized as the geographies of brands and branding. They provide assets and resources for the efforts of actors in trying to summon up and articulate meaning and value in certain market contexts. This is not to claim that the meaning and value of brands and branding is only made up of geographical associations, but it is to argue that the attributes and characteristics of brands and branding have inescapable geographical associations with which actors have to work. Origination distinguishes geographical associations of differing kinds, extents and characters. As attachments to spaces and places they are not only singular, fixed and stable spatial ties that determine the geographies of brands and branding. Geographical associations can be multiple, fluid and unstable in their different form, degree and nature. Distinguishing the variegation of geographical associations in brands and branding demonstrates how actors shift originations across spatial scales and within relational circuits and networks. Origination is not only bounded and scalar; it engages tensions and accommodations with unbounded and relational geographical associations. Geographical associations are only framed and fixed temporarily in particular times and spaces by specific actors in ephemeral market settings. This openness and fluidity in geographical associations makes them pliable and functional for at least some brand and branding actors. Geographical differentiation often exists between where the different activities in the spatial circuits of branded goods and services commodities are undertaken. As the empirical analyses of Newcastle Brown Ale, Burberry and Apple demonstrate, geographical associations can be made of certain kinds to particular places for specific sorts of activities within broader spatial circuits. Where the design, assembly, advertising, retailing and patenting activities in production, circulation, consumption and regulation take place can matter – not always and everywhere – to the meaning and value of the branded goods and services commodities in spatial and temporal market settings.

Origination reveals how the meaning and value of some brands and their branding can be sustained by actors in specific markets even with what appear to be a lack of material or weaker geographical associations to particular places. Actual and/or perceived understandings about brands and their branding may or may not be supported by any empirical substance because of the ways in which brand and branding actors construct, connote or even confuse – deliberately or otherwise – authentic and fictitious geographical associations.

Second, origination revealed that meaningful and valuable geographical associations constructed and cohered in brands and branding by participant actors are inherently unstable in the context of dynamic economic rationales *and* cultural-economic tendencies. Logics of accumulation, competition, differentiation and innovation animate and compel actors in spatial circuits. They force the ongoing and restless disruption of fixes of meaning and value accomplished by actors seeking commercial growth in specific periods and particular places. Echoing David Harvey's (1996: 293) 'conditional permanancies' continually in flux, a brand and its branding provide only temporary coherence, shape and form for the geographical associations that actors deploy in trying to create meaning and value. The evolving character and complexity of geographical imaginaries and places over time too makes them difficult and slippery for brand and branding actors to pin down in meaningful and valuable ways. This reading of the constructed nature of brands and branding challenges and qualifies accounts that suggest there is some kind of timeless, spaceless, placeless and fixed essence to brands and their branding. David Aaker (1996: 68), for example, refers to the 'central, timeless essence of a brand'. Michel Chevalier and Gérald Mazzalovo (2004: 98) claim 'a brand can have meaning beyond the products themselves and the advertising slowly developed' as the 'vocabulary was enriched with concepts such as the "essence", "*raison d'être*", "consciousness", "soul", and "genetic code" of a brand'. Origination questions and moves beyond any such deterministic view of brands and branding. The social construction of brands and branding by people in places inescapably imbues them with meaningful and valuable geographical associations. Brand and branding actors wrestle constantly with the constantly unfinished task of origination across and between territorial scales *and* relational networks. Logics of accumulation, competition, differentiation and innovation unsettle, disturb and undermine meaningful and valuable origination fixes in spatial circuits of meaning and value. The concept and theory of origination and socio-spatial biography method aimed to understand and explain this enduring phenomenon.

Third, origination illuminated the ways in which social and spatial inequalities get reproduced by actors constructing geographical associations in brands and branding in geographically uneven ways. Spatial differentiation of economy, society, polity, culture and ecology underpins the dynamics of brand and branding. Actors in spatial circuits are compelled by the dynamics of accumulation, competition, differentiation and innovation to search for, create, exploit and (re)produce economic and social disparities and inequalities over space and time. Demonstrated in the empirical analyses of Newcastle Brown Ale, Burberry and Apple, this geographical differentiation is manifest in the spatial circuits producing, circulating, consuming and regulating brands and branding. It is evident in the changing material connections of breweries, factories, offices, subcontractors and service centres and the investments, jobs and incomes they involve. It is found in the symbolic and discursive geographical associations constructed to particular spaces and places that shape perceptions of the capability, role and reputation of Newcastle upon Tyne and England, Britain and 'Britishness', and Silicon Valley and California.

As a conceptual, theoretical and analytical framework, origination provides the means of lifting the 'mystical veils' (Greenberg 2008: 31) woven by brand and branding actors within spatial circuits as they try to construct and cohere meaning and value in certain markets times and spaces. Origination connects with the calls of David Harvey (1990) and Michael Watts (2005) to get behind and unveil the commodity fetish and trace the relationships between commodities and uneven geographical development. It goes further by grounding them in the contemporary context of branded 'cognitive-cultural capitalism' (Scott 2007: 1466). Against the critique of 'de-fetishization', origination can contribute several points. First, whether the commodity fetish is conceived in single, double (Cook and Crang 1996) or more complex, layered and multiple forms (Smith and Bridge 2003; Castree 2001), origination conceptualizes and theorizes how, why and where actors construct geographical imaginaries through branded commodities and their underlying socio-spatial relations. Second, rather than privileging any particular source (Jackson 1999), origination is receptive to the knowledges of actors involved in brands and branding throughout the *full* spatial circuit of production, circulation, consumption and regulation. It is open to the agency of public, private, civic and hybrid social groups, individuals and institutions (See Table 2.10). Last, origination underlines the continued – perhaps even reinforced – importance and relevance of the commodity fetish in branded 'cognitive-cultural capitalism' (Scott 2007: 1466). While it gets behind the veils of brands and branding, origination does not claim this will necessarily encourage and mobilize actors to work against any revealed social and spatial inequalities and uneven geographical development. It does not argue that commitments and responsibilities already exist that simply require some kind of activation by being revealed as the realities of branded commodities and capitalism (cf. Barnett *et al.* 2005: 24). The political and cultural economy of the geographies of brands and branding underpinning origination seek to conceptualize, theorize and analyse, and to enhance our understanding and explanation and to inform our praxis.

Fourth, the methodological and analytical approach to empirical research explored the worth of origination and geographical associations in explaining the agency of actors attempting to create and fix meaning and value in brands and branding over space and time. Connecting political and cultural economy, socio-spatial biography addressed complexity, diversity and variety in cultural construction and grasped the systematizing logics, rationales and tendencies of accumulation, competition, differentiation and innovation in contemporary branded capitalism. Socio-spatial biography engaged the ways in which brand and branding actors tap into particular geographical and temporal imaginaries for selective reworking, sometimes blurring the authentic with the fictitious. The empirical analysis addressed Newcastle Brown Ale's imported British or English ale from the 1990s and 2000s, Burberry's modernized 'Swinging London' from the 1960s and Apple's revolutionary high-tech innovation from Silicon Valley in California from the 1970s. As complex and multi-faceted social and spatial entities, any place is not simply reducible to a brand even though branding actors often seek such reductionist and shorthand simplification. The method, research design and analytical framework introduced here to engage origination have potential wider merit. They need to be explored and tested in new and comparative empirical cases and contexts. Especially interesting would be, first, work to examine goods, services and knowledges with different and potentially challenging origination(s). Learning from the empirical ambition of the commodity 'following' work (e.g. Cook *et al.* 2006), further studies might explore branded commodities whose geographical associations and origination may appear – at least at first glance – to be less obvious: credit cards, headache tablets, insurance, internet commerce, mobile communications, pet food, shampoo, supermarket own-brands, transportation and so on (Pike 2011d). Second, international comparative research could explore how the same brands are originated by different actors in different ways in different spatial and temporal market contexts. This is what Daniel Miller (2008: 23) terms the 'diversity of branding' revealed by 'a more nuanced ethnography of contemporary branding as a material practice'.

Last, examining the origination of brands and branding highlights the relationships and tensions between political *and* cultural economy. Origination theorizes how actors use geographical associations to constitute and originate brand meaning *and* value. Each needs to be considered together to explain how attempts are made to cohere and stabilize them through the agency of actors in spatial circuits. The focus needs to extend beyond just production and consumption to circulation and regulation. Just focusing on production and/or consumption risks providing only a limited account. Integration within the full circuit with circulation (Hughes 2006) and regulation (Willmott 2010) is critical to a deeper understanding and explanation. Socio-spatial biography reveals how the 'spirit' or 'personality' (Molotch 2002: 666) of particular places can be appropriated in geographical associations by producers, circulators, consumers and regulators. A partial view results if such geographical associations are considered only as cultural constructions – the replacement of use and exchange value by sign value (Edensor and Kothari 2006: 332) – without connection to the commercially inspired 'production of difference' (Dwyer and Jackson 2003) and political-economic disciplines of 'costs and cash' (Sayer 1997: 22). Origination integrates political economy insights into the rationales animating and disrupting actors in spatial circuits of value with a stronger grasp of identity and meaning construction. It provides a way of engaging the creation of economic value *and* 'manufacture of meaning' (Jackson *et al.* 2007) within spatial circuits. Dialogue between political *and* cultural economy is promising because 'even though the outputs of the cultural economy have high symbolic meaning, this system of production is also regulated by the down-to-earth goal-oriented strategies of firms and workers structured as they are by a significant concern for monetary returns to investment and effort in the context of market forces' (Scott 2010: 120; see also Hudson 2008).

Origination encourages reflection upon how central spatial concerns in the geographies and dynamics of political and cultural economy are understood and explained. For (sub-)disciplines with interests in such geographical issues – especially marketing and sociology – the focus upon the role and importance of brands and branding and the origination of their geographical associations by actors in spatial circuits provides a means of conversing between political and cultural economy. Rooted in the product, the service, the firm, the industry and, more recently, the global production network and value chain as its 'units of analysis' (*Economic Geography* 2011: 113), economic geography in particular can benefit from such an endeavour in which brands and branding provide a complementary point of departure and focus of conception, theorization and analysis. Bridging and linking insights from across (sub-)disciplinary research in geography and engaging accounts from wider social science, origination and its approach to explaining and researching the geographies of brands and branding can further work and understanding across disciplines (Pike 2011d). An aspiration for initiating cross-disciplinary dialogue (Peck 2012), establishing 'trading zones' (Barnes 2006), negotiating 'bypasses' and 'risky intersections' (Grabher 2006), even contributing to 'post-disciplinarity' (Sayer 1999), underpins such a project.

The politics of origination

Building upon origination as a conceptual, theoretical and analytical framework prompts reflections on critical, political and normative issues. Interpreting geographical associations deployed by actors within spatial circuits in their efforts to construct and stabilize meaning and value in spatial and temporal market settings affords a means to consider the politics and limits of branded cognitive-cultural capitalism. In asking normative questions about 'what kind of brands and branding and for whom?' (Pike 2011), origination contributes to tackling difficult questions about any 'radical' and/or 'sustainable' politics of consumption (Cook *et al.* 2007). It scrutinizes how branded goods and services commodities might be associated geographically in more developmental and progressive ways for people and places.

Origination furthers existing critical work illuminating the limits of branded capitalism. Counter-critiques emerged in the wake of Naomi Klein's *No Logo* in the early 2000s. For example, Michel Chevalier and Gérald Mazzalovo's (2004: 3) *Pro Logo* and its claim that:

> Brands exist and are neither good nor evil in themselves. They can be criticized, but calling for their abolition is absurd. They are and will remain an essential tool of marketing, international competition, and contemporary social life. It's impossible to imagine that supermarkets would suddenly begin selling exclusively generic products. If that were to happen, the need for differentiating these products would immediately arise – and brands would reappear, or else the store's name would take their place. If fact, there is no such thing as a world without brands.

A decade on, political readings of brands and branding remain divided, even polarized. Martin Kornberger (2010: 205) describes the opposing schools of thought split between, first, the 'liberal tradition' and its 'ideology of free markets' where the 'brand plays an important role in the defence of free market principles: brands create accountability, loyalty and wealth for everybody. In other words, brands liberate'. And, second, the 'more critical tradition' that 'interprets brands as part of the problem not the solution. Brands are the lubricants of free-markets in which the consumer mentality bulldozes all other forms of life. Brands are the avant-garde of the capitalist quest for world domination, spearheading an invasion of culture and privacy.' Such differing views both have to address and reflect upon the emergent and increasingly apparent limits to brands and branding. Instead of marks of quality and consumer protection, branded goods and services market saturation is causing brand anxiety, disillusion, fatigue, 'blindness' (Klein 2000: 13) and consumer indebtedness. Attempts at competitive emulation of Veblenian 'conspicuous consumption' by people means 'Once others gain access to what you have, new stuff has to be acquired in an endless cycle of unhappy waste' (Molotch 2005: 4). Consumer sovereignty is revealed as illusory in branded cognitive-cultural capitalism as Raymond Williams' (1980) 'magic system' of advertising weaves fantasies around what corporations decide to supply as the market 'choices' of discerning consumers (Hudson 2005: 70). Hyperactive brand and branding by actors in spatial circuits risks panic over branding of goods, services, lifestyles, spaces and places as 'firms hurl inflated ad budgets and the kitchen sink of signifiers into frantic efforts to stand out in image markets' (Goldman and Papson 2006: 328–9). Brands and branding are foisted upon bewildered consumers ever more pervasively in competitive markets, updating them ever more rapidly for faster capital turnover (Harvey 1989), compensating for increased failures and diminishing returns to capital from ever higher levels of brand promotion investment (Riezebos 2003). Actors aim to stave off what Naomi Klein (2000: 118) saw as 'the nightmare moment when branded products cease to look like lifestyles or grand ideas and suddenly appear as the ubiquitous goods they really are'.

Origination reveals the temporary and unstable nature of geographical associations used by actors trying to create and fix meaning and value. Brand and branding geographies are fragile and vulnerable to fashion vagaries, commercial rivalry and displacement. Counterfeit 'knock-offs' have grown internationally (Molotch 2005). Consumer dissent – such as Neil Boorman's (2007) 'bonfireofthebrands.com' blog and book – has increased against paying the premium price of the 'brand tax' (Riezebos 2003: 24). Brand fragility and collapse has resulted in the disappearance of – albeit for different reasons – formerly established brands including Accenture, Barings Bank, Enron, Lehman Brothers, Marconi, Norwich Union and Worldcom. Amidst the apparent normalization of brands and branding in consumer society, brand-based activism has mobilized around their visibility and value. It has tried to target, resist and subvert brands as symbols of capitalist globalization and to exploit their fragile and unstable nature. High profile brands have become visible targets of political-economic activism. Examples include: anti-capitalist and green movement direct action against McDonalds and Starbucks as branded symbols of capitalist globalization (Bové and Dufour 2002; Klein 2000); the anti-sweatshop campaigns' focus upon Gap and Nike's international outsourcing (Ross 2004); the questioning of the Burberry brand's 'Britishness' by the 'Keep Burberry British' campaign in seeking to prevent a factory closure in Wales following international subcontracting (Chapter 5); and the controversy concerning the contribution of Apple to growth, jobs, recovery and taxes in America (Chapter 6). But critical and political engagement with brands and branding actors in spatial circuits is not a straightforward or unproblematic task. Whether approached as entry points or diversions, brands are politically ambiguous. As their substantive and geographical reach is extended throughout economy, society, culture, polity and ecology, politics has lagged in 'demanding a citizen-centered alternative to the international rule of the brands' (Klein 2000: 246). Political difficulties are multiple: the lack of reflexivity and uncertainty about brand-based activism's meanings (Littler 2005); the socially and spatially uneven extent of citizens' political consciousness about brand provenance (Ross 2004) and whether they care or not about the origination of the goods and services they consume; the emergence of 'brand-based activism' as 'the ultimate achievement of branding' (Klein 2000: 428); the narrow focus upon 'designer injustices' (Klein 2000: 423); the marketing of resistance symbols back to brand conscious dissenters; and the co-option of resistance movements by brands and branding (Huish 2006). Indeed, Martin Kornberger (2010: 23) even claims that 'The anti-brand manifesto *No Logo* is a brand, just as much as the anti-advertisers and sub-vertisers from *Adbusters* are a brand'. This point has been conceded by Naomi Klein (2010: xvi) in 'the brand that I had accidentally created: No Logo' and the coping strategy of breaking brand and branding rules 'whenever the opportunity arose'.

Origination affords a way to consider the progressive political potential of the inescapable geographical associations deployed by actors trying to create and fix meaning and value in brands and branding. The geographies of brands and branding can provide a 'non-abstract starting point' (Klein 2000: 356) to frame political questions of social and spatial justice and distribution concerning who and where benefits or loses from particular kinds of geographical association and origination. This is not a call simply to deconstruct complexity, diversity and plurality in commodity brand and branding geographies (e.g. Cook *et al.* 2007). That approach risks 'retreating into the discovery of fragments and contingencies' (Perrons 1999: 107). Scrutinizing the origination of brands and branding through the culturally sensitive political-economic construction of their socio-spatial biographies – undertaken here for Newcastle Brown Ale, Burberry and Apple – foregrounds their connections to politics and uneven geographical development. It identifies the agents involved and questions the geographical associations of brands and branding and what they mean for territorial development. Rather than dictating meaning (cf. Cook *et al.* 2007), it provides explanation and focuses reflection upon alternative geographical imaginaries that can put brands and branding in their place; deflating the hype and reminding of their historical role as marks of identification and quality. Origination and the geographical associations of brands and their branding analyse their connections to people and places. Such geographical imagination opens up avenues for deliberation and action and suggests opportunities for regulatory agency.

Origination and geographically uneven development

Connecting origination to broader debates about the commons (Harvey 2006) and new, more distributed, deliberative and socialized forms of public ownership (Cumbers 2012) raises a central political issue: whether or not people in places could or should have a form or degree of social ownership and/or control of brands where actors are appropriating attributes and characteristics of 'their' place as a source of differentiated meaning and value through branding practices. Engaging this question requires a shift in thinking to view especially brands whose originated meaning and value depend upon strong, deep and enduring geographical associations as collective and public rather than individual and private assets embedded in place. Such brands might then be owned, controlled and managed by some form of social organization. This could be civic associations, cooperatives, social groups or mutual societies. National and supranational regulatory support could help sustain brand quality, encourage collective innovation and contest detachment from place (see Morgan *et al.* 2006). Thinking and practising origination in this way exists in the current use of Geographical Indications (GIs). As regulatory devices, GIs contrast the securitization of geographical associations in brands and branding through trademarks that convert attributes of place and local knowledges into property, rendering such associations private, tradeable and vulnerable to de-localization (Morgan *et al.* 2006). The alternative view of a public and collective commons of assets and attributes in place is antithetical to the 'accumulation by dispossession' (Harvey 2006) and enclosure of the commons advanced by legally entrenched and exclusive private property relations in branded cognitive-cultural capitalism. But it can prompt discussion, dialogue and debates about what kinds of 'development' are sought regionally and locally (Pike *et al.* 2007). It may encourage reflection too upon the erosion of longstanding and traditional forms of identity based on class, kinship and place by image-saturated and internationalized goods and services brands and their branding (Wengrow 2008). Public consciousness of private enclosure and the trademarking of the cultural commons by what David Bollier (2005) calls 'brand name bullies' is growing, albeit unevenly, and fuelling contestation. Sportswear brand Nike, for example, was ordered to pay £300 000 in a settlement after its appropriation of the London Borough of Hackney's corporate logo in its sportswear was judged illegal following the use of images of celebrity football games on Hackney Marshes in one of its TV advertising campaigns (*Hackney Today* 2006).

Given the limits, political concerns and uneven geographical development inscribed in branded capitalism, origination prompts fresh thinking about how actors in spatial circuits could be encouraged and facilitated geographically to associate their brands and branding with especially less prosperous, peripheral and weaker places across the world in more developmental ways. Returning to the case of the 'national' origination of Burberry, interesting wider reflections on the origination of brands, branding and territorial development were triggered by the closure of its factory in Treorchy in Wales. Over 300 jobs were lost leaving an enduring sense of devastation in this small and tight-knit local community as former workers still struggled to secure employment several years later (Wales Online 2011). Rhys David (2007: 1) articulated an anxiety over economic prospects familiar to many former industrial and peripheral regions internationally. In a world where the brands and branding of goods and services commodities are increasingly important to demonstrate who or what a place is good at which is internationally competitive and resonant, he interprets such places as having 'only half a brand' based upon their industrial heritage, talented individuals and/or ability to host short-term spectacle events.

Bemoaning the lack of nationally framed and originated 'Welshness' capable of delivering branding benefits to the national territory of Wales within the United Kingdom through more direct connections with consumer markets, Rhys Davis (2007: 1) argued that:

> The absence of indigenous Welsh companies projecting their own brand also makes it much more difficult to adapt and survive in a world where globalisation is leading to the concentration of production in the lowest cost centres, China and India ... [which] are going to dominate world manufacturing for the foreseeable future. You do not, however, have to *make* products any more but, any economy with ambition does need to be involved in *organising* their design, manufacture, distribution and sale, if it is to share in the benefits of globalisation and not always be its victim. These higher value-added areas are where the jobs are increasingly being created in advanced economies.

Emphasizing the issue of brand ownership and control, he went on to summarize the development problem for such territories: 'If you lack the locally-owned company with its brand in the first place, you are not in a position to outsource your production. Companies – such as Burberry which is moving manufacturing from Wales to China – move out and leave no trace behind because they have their headquarters elsewhere and will organise their production in overseas countries from that location. The higher added value jobs remain in the UK but not in Wales, because we never "owned" the brand in the first place' (David 2007: 1).

In many ways this analysis is nothing new and the concerns it raises extend internationally beyond the specific territory of Wales. It echoes longstanding debates about ownership, external control, the 'branch plant economy' and the shifting roles and positions of places within the international division of labour. But, importantly, it situates these enduring issues in the context of contemporary concerns with 'creativity', higher value-added and 'knowledge-intensive' economic activities as well as the pervasive and influential worlds of brands and branding. This is where origination can help to open up some interesting questions about the potential for territorial development. Origination in the labelling of branded goods and services has become more attuned in representing the geographically differentiated ways in which economic activities are organized internationally. Some actors are now explicitly recognizing and articulating in their brands and branding the meaning and value of where the brand was from, where it was thought up and designed, where its owner was based, where – if it was a service – it was delivered from and, for goods, where it was made (Chapters 3 and 6). Origination broadens the analytical focus of territorial development to capture the potential benefits of economic activities from other moments in spatial circuits of production, circulation, consumption and regulation. It stimulates thinking about the extent and nature of the ways in which regional and local economies are involved – or not – in what Rhys David referred to as the *organization* of economic activities? Origination opens up a much broader array of things that places could do which could be used to shape their development: innovation, design, prototyping, testing, manufacture, assembly, marketing, distribution, sale, servicing and regulation. Moving beyond the problematic brands and branding of spaces and places, it asks how could places get recognized and acknowledged by specialized capability, reputation and renown as at least one of *the* places to organize and/or undertake such activities as a way of generating sustainable paths of territorial development. In a context where economic activities are increasingly carved up, distributed and connected internationally within the kinds of global value chains discussed in Chapter 3, origination encourages thinking about the role, position and reputation of less prosperous, peripheral and weaker places across the world within this broader set of economic activities. Origination sparks thinking about how individuals and institutions can attempt to shape their evolution over time in ways beneficial for their aims, aspirations and conceptions of territorial development.

In raising such issues there are several clarifications to make. First, this is not a claim that the meaning and value of origination for particular places is set in stone. While the trajectories of past development paths can be important in patterning future routes, they need not be inevitable processions determined by their histories alone (MacKinnon *et al.* 2009). Similarly, this argument is not a call to abandon the production activities of manufacture and assembly in a blind search only for 'higher-value added' and service-oriented activities. This strategy would be especially problematic for places that have built a capability, reputation and renown as well as jobs and incomes for people from such activities. Production remains an integral part of spatial circuits of value and meaning. The goods and services and the parts and activities from which they are made and provided have to be produced somewhere. As discussed in Chapter 3, both tangible and intangible attributes in branded goods and services commodities are becoming more important sources of meaning and value in spatial and temporal market contexts. On- or re-shoring of assembly and manufacturing activities back to western economies is occurring (Christopherson 2013). Even Apple's internationally outsourced business model is being questioned and its managers are bringing some manufacturing back to America (Chapter 6). The argument here is that production needs to be situated within its broader spatial circuit with circulation, consumption and regulation more thoroughly and imaginatively to address its origination and potential for territorial development. Moreover, the polarization of labour markets between high and low level occupations apparent in transitions to more knowledge-intensive and higher value-added economic activities could benefit from some rounding and progression ladders through the creation of medium level semi-skilled and intermediate jobs (Turok 2011). This argument is also not suggesting that peripheral places should get more actively involved in the artificial construction of geographical associations in a bid to create meaning and value. North Northamptonshire Development Corporation, for example, mounted a £1.3 million marketing campaign in an effort to rebrand Northamptonshire as 'North Londonshire' and emphasize its proximity and connections to London and the greater south east to attract investment, jobs and residents (*BBC News* 2010). This kind of construction is risky as knowledgeable consumers can react negatively and those implicated can dissent as the Northamptonshire residents' anti-brand Facebook campaign demonstrates.

Going with the grain of existing, even dormant and underutilized assets and skills can provide a way to build or rebuild capability, reputation and renown for particular activities such as design, testing and production in new and growing markets (Dawley 2014). Whether codified as brands or not, origination asks the question of how such competences can be readily identified in *and* with places. It even prompts consideration of whether origination can act to communicate and project positive images and reputation internationally – perhaps boosting self-esteem, pride and confidence in a place – and provide some positive stimulus to territorial development regionally and locally (Chapter 7). Rhys David (2007: 1) argues much the same for Wales in his call for 'creating more and making more out of strong vibrant businesses and brands that can be readily identified as Welsh as markers and envoys of reputation in the wider world'. Even the Conservative Chancellor of the Exchequer in the UK Coalition Government, George Osborne (2011), appears to recognize origination in stating in a recent Budget Speech:

> We are only going to raise the living standards of families if we have an economy that can compete in the modern age. So this is our plan for growth. We want the words: 'Made in Britain', 'Created in Britain', 'Designed in Britain', 'Invented in Britain'. To drive our nation forward. A Britain carried aloft by the march of the makers. That is how we will create jobs and support families. We have to put fuel into the tank of the British economy.

Asking questions about the origination of what less prosperous, peripheral and weaker places are and/or can be good at, then, opens up new issues and debates about their development aspirations, potential and prospects. The context of material challenges to existing forms of economic, social, cultural, political and ecological organization through climate change, financialization, resource shortages and social inequality underlines this need for dialogue and deliberation about the origination of brands and branding in territorial development.

References

Aaker, D. A. 1996. *Building Strong Brands*. New York: The Free Press.
Aaker, J. L. 1997. 'Dimensions of Brand Personality.' *Journal of Marketing Research*. 34, 347–356.
Adbusters. 2012. 'Meme Wars: The Creative Destruction of Neoclassical Economics.' Accessed 26 November 2014. www.adbusters.org.
Agnew, J. 2002. *Place and Politics in Modern Italy*. Chicago, IL. University of Chicago Press.
Allen, J. 2002. 'Symbolic Economies: The "Culturalization" of Economic Knowledge.' In *Cultural Economy: Cultural Analysis and Commercial Life*, edited by P. Du Gay and M. Pryke, 39–58. London: Sage.
Amin, A. 2004. 'Regions Unbound: Towards a New Politics of Place.' *Geografiska Annaler*. 86 B 33–44.
Anholt, S. 2006. *Competitive Identity: The New Brand Management for Nations, Cities and Regions*. Basingstoke: Palgrave MacMillan.
Anholt, S. and Hildreth, J. 2004. *Brand America: The Mother of All Brands*. London: Cyan Books.
Appadurai, A. 1986. 'Introduction: Commodities and the Politics of Value.' In *Social Life of Things*, edited by A. Apparadurai, 3–63. Cambridge: Cambridge University Press.
Apple. 2012. '2012 10-K/A.' Accessed 26 November 2014. http://www.apple.com/investor/.
Apple. 2013. 'Creating Jobs Through Innovation.' Accessed 26 November 2014. http://www.apple.com/about/job-creation/.
Aronczyk, M. and Powers, D. 2010. *Blowing Up the Brand*. New York: Peter Lang.
Aronczyk, M. 2013. *Branding the Nation: The Global Business of National Identity*. Oxford: Oxford University Press.
Arvidsson, A. 2005. 'Brands: A Critical Perspective.' *Journal of Consumer Culture*. 5, 2, 235–258.
Arvidsson, A. 2006. *Brands: Meaning and Value in Media Culture*. London and New York: Routledge.
Ashworth, G. J. and Kavaratzis, M. 2010. Eds. *Towards Effective Place Brand Management: Branding European Cities and Regions and Regions*. Cheltenham, UK and Northampton, MA, USA: Edward Elgar.

Origination: The Geographies of Brands and Branding, First Edition. Andy Pike.
© 2015 John Wiley & Sons, Ltd. Published 2015 by John Wiley & Sons, Ltd.

Ashworth, G. and Voogd, H. 1990. *Selling the City: Marketing Approaches in Public Sector Urban Planning*. London: Belhaven Press.

Askegaard, S. 2006. 'Brands as a Global Ideoscape.' In *Brand Culture*, edited by J. Schroeder and M. Salzer-Mörling, 91–102. New York: Routledge.

Banks, G., Kelly, S., Lewis, N. and Sharpe, S. 2007. 'Place "From One Glance": The Use of Place in the Marketing of New Zealand and Australian Wines.' *Australian Geographer.* 38, 1, 15–35.

Barham, E. 2003. 'Translating Terrior: The Global Challenge of French AOC Labelling.' *Journal of Rural Studies.* 19, 127–138.

Barnes, T. 2006. 'Lost in Translation: Towards an Economic Geography as Trading Zone.' In *Denkanstöße zu einer anderen Geographie der Ökonomie*, edited by C. Berndt and J. Glückler 1–17. Bielefeld: Transcript.

Barnes, T., Peck, J., Sheppard, E. and Tickell, A. 2007. 'Methods Matter.' In *Politics and Practice in Economic Geography*, edited by A. Tickell, T. Barnes, J. Peck and E. Sheppard. Thousand Oaks, CA: Sage.

Barnett, C. Cloke, P., Clarke, N. and Malpass, A. 2005. 'Consuming Ethics: Articulating the Subjects and Spaces of Ethical Consumption.' *Antipode.* 37, 23–45.

Barrientos, S., Mayer, F., Pickles, J. and Posthuma, A. 2011. 'Labour Standards in Global Production Networks: Framing the Policy Debate.' *International Labour Review.* 150, 3–4.

Barry, A. and Slater, D. 2002. 'Introduction: The Technological Economy.' *Economy and Society.* 31, 2, 175–193.

Bass, F. and Wilkie, L. 1973. 'A Comparative Analysis of Attitudinal Predictions of Brand Preference.' *Journal of Marketing Research.* 10, 262–269.

Batchelor, A. 1998. 'Brands as Financial Assets.' In *Brands: The New Wealth Creators*, edited by S. Hart and J. Murphy, 95–103. Basingstoke: MacMillan.

Bauer, R. A. 1960. 'Consumer Behaviour as Risk Taking.' In *Dynamic Marketing for a Changing World*, edited by R. S. Hancock. Boston, MA: American Marketing Association.

Bauman, Z. 2007. *Consuming Life*. Cambridge: Polity.

BBC. 2011. 'Chinese Authorities Find 22 Fake Apple Stores.' BBC Technology News, 12 August. Accessed 26 November 2014. http://www.bbc.co.uk/news/technology-14503724.

BBC News. 2007. 'Row over 'Burberry Hate Campaign." 4 February 2007. Accessed 26 November 2014. http://news.bbc.co.uk/1/hi/wales/mid/6329703.stm.

BBC News. 2010. "North Londonshire' Label for Northamptonshire Attacked.' 3 March. Accessed 26 November 2014. http://news.bbc.co.uk/1/hi/england/northamptonshire/8548647.stm.

BBC Tyne. 2004. 'Gateshead Brown Ale. BBC Tyne Features.'

Belk, R. W. and Tumbat, G. 2005. 'The Cult of Macintosh.' *Consumption, Markets and Culture.* 8, 3, 205–217.

Bell, F. 2013. "Reputation of Place – Literature Review." Unpublished Paper, CURDS: Newcastle University.

Bellini, N. 2011. *Researching Place Branding: The New Agenda*. Paper for the Regional Studies Association Conference, Newcastle upon Tyne, April.

Bennison, B. 2001. 'Drink in Newcastle.' In *Newcastle Upon Tyne: A Modern History*, edited by R. Colls and B. Lancaster, 167–192. Chichester: Phillimore.

Beverland, M. B. 2009. *Building Brand Authenticity: 7 Habits of Iconic Brands*. New York: Palgrave MacMillan.

Bilkey, W. J. and Nes, E. 1982. 'Country-of-Origin Effects on Product Evaluations.' *Journal of International Business Studies.* 8, 1, 89–99.

Bollier, D. 2005. *Brand Name Bullies: The Quest to Own and Control Culture*. Hoboken, NJ: Wiley.

Boorman, N. 2007. *Bonfire of the Brands: How I Learnt to Live Without Labels*. Edinburgh: Canongate.
Borrus, M. and Zysman, J. 1997. *Wintelism and the Changing Terms of Global Competition: Prototype of the Future?*, Working Paper 96B February, DRUID. Aalborg, Denmark: Aalborg University.
Boschma, R. 2005. 'Proximity and Innovation: A Critical Assessment.' *Regional Studies*. 39, 1, 61–74.
Bové, J. and Dufour, F. 2002. *The World Is Not For Sale: Farmers Against Junk Food*. London and New York: Verso.
Bowers, S. 2006. 'The Beer Necessities.' *The Guardian*, London, 26 February.
Braithwaite, D. 1928. 'The Economic Effects of Advertising.' *Economic Journal*. 38, 16–37.
Brand Republic. 2007. 'Burberry Unveils Spring Collection with Beaton-inspired Ads.' 5 January. London: Brand Republic.
Breward, C. and Gilbert, D. 2006. *Fashion's World Cities*. London: Berg.
Bryant, C. 2007. Royal Warrants of Appointment. *House of Commons Hansard Debates*, Col426WH, 23 January.
Bryant, R. 2014. 'The Fate of the Branded Forest: Science, Violence and Seduction in the World of Teak.' In *The Social Life of Forests*, edited by S. Hecht, K. Morrison and C. Padoch. Chicago, IL: University of Chicago Press.
Buck Song, K. 2011. *Brand Singapore: How Nation Branding Built Asia's Leading Global City*. Singapore: Marshall Cavendish.
Bulkeley, H. 2005. 'Reconfiguring Environmental Governance: Towards a Politics of Scales and Networks.' *Political Geography*. 24, 875–902.
Bunting, M. 2001. 'Clean Up.' *The Guardian*, 6 October, London.
Burberry. 2005. *Annual Report and Accounts 2004/05*. London: Burberry.
Burberry. 2006. *Annual Report and Accounts 2005/06*. London: Burberry.
Burberry. 2007. 'About Burberry.' Accessed 26 November 2014. http://www.burberry.com/AboutBurberry/History.aspx
Burberry. 2008. *Annual Report and Accounts 2007/08*. London: Burberry.
Burberry. 2009. *Annual Report and Accounts 2008/09*. London: Burberry.
Burgess J. A. 1982. 'Selling Places: Environmental Images for the Executive.' *Regional Studies*. 16, 1–17.
Buzzell, R. D., Quelch, J. A. and Bartlett, C. A. 1995. *Global Marketing Management: Cases and Readings*. Reading, MA: Addison-Wesley.
Callan, E. 2006. 'Burberry Aims for Anglophiles in US Heartland.' *The Financial Times*, 8 July, London.
Cadwalladr, C. 2012. 'The Hypocrisy of Burberry's "Made in Britain" Appeal.' *The Guardian*. 16 July, London.
Callon, M. 2005. 'Why Virtualism Paves the Way to Political Impotence: A Reply to Daniel Miller's Critique of The Laws of the Markets.' *Economic Sociology – European Electronic Newsletter*. 6, 2, 3–20.
Callon, M., Méadel, C. and Rabeharisoa, V. 2002. 'The Economy of Qualities.' *Economy and Society*. 31, 2, 194–217.
Campbell, C. 2005. 'The Craft Consumer: Culture, Craft and Consumption in a Postmodern Society.' *Journal of Consumer Culture*. 5, 1, 23–42.
Carlsberg. 2012. 'Annual Report.' Accessed 26 November 2014. http://www.carlsberggroup.com/investor/downloadcentre/Documents/Annual%20Report/Carlsberg%20Breweries%20Annual%20Report%202012.pdf.

Carrell, S. 2012. 'Island Crofters' Cloth Joins the Fashion Mainstream.' *The Guardian*, 10 November, London.

Casson, M. 1994. 'Brands: Economic Ideology and Consumer Society.' In *Adding Value – Brands and Marketing in Food and Drink*, edited by G. Jones and N. Morgan, 41–58. London: Routledge.

Castree, N. 2001. 'Commodity Fetishism, Geographical Imaginations and Imaginative Geographies.' *Environment and Planning*. 33, 1519–1525.

Chakrabortty, A. 2011. 'Why Doesn't Britain Make Anything Anymore?' *The Guardian*, 16 November, London.

Chesbrough, H. 2003. *Open Innovation: The New Imperative for Creating and Profiting from Technology*. Cambridge, MA: Harvard Business School Press Books.

Chevalier, M. and Mazzolovo, G. 2004. *Pro Logo*. New York: Palgrave MacMillan.

Christopherson, S. 2013. 'The Regional Advantage: What the Manufacturing Location Calculus Implies For the NE Local Enterprise Partnership Independent Economic Review.'

Christopherson, S., Garretsen, H. and Martin, R. 2008. 'The World is Not Flat: Putting Globalization in its Place.' *Cambridge Journal of Regions, Economy and Society*. 1, 3, 343–349.

Clark, G. L. 1998. 'Stylized Facts and Close Dialogue: Methodology in Economic Geography.' *Annals of the Association of American Geographers*. 88, 1, 73–87.

Coe, N. M., Hess, M., Yeung, H. W.-C., Dicken, P. and Henderson, J. 2004. '"Globalizing" Regional Development: A Global Production Networks Perspective.' *Transactions of the Institute of British Geographers NS*. 29, 468–484.

Collins Concise Dictionary Plus. 1989. *Collins Concise Dictionary Plus*. London and Glasgow: Collins.

Comhairle nan Eilean Siar. 2003. *Regional Accounts*. Stornoway: Comhairle nan Eilean Siar.

Competition Commission. 1989. 'Elders IXL Ltd and Scottish and Newcastle Breweries PLC: A Report on the Merger Situations.' Competition Commission: London.

Connelly, M. 2012. 'Poll finds Confusion on Where Apple Devices are Made.' *The New York Times*, 25 January. Accessed 26 November 2014. http://www.nytimes.com/2012/01/26/business/poll-on-iphone-and-ipad-finds-consumer-confusion-on-apples-manufacturing.html.

Cook, I. and Crang, P. 1996. 'The World on a Plate: Culinary Culture, Displacement and Geographical Knowledges.' *Journal of Material Culture*. 1, 131–153.

Cook, I. and Harrison, M. 2003. 'Cross Over Food: Re-materializing Postcolonial Geographies.' *Transactions of the Institute of British Geographers NS*. 28, 296–317.

Cook, I. *et al.* 2006. 'Geographies of Food: Following.' *Progress in Human Geography*. 30, 5, 655–666.

Cook, I., Evans, J., Griffiths, H., Morris, R. and Wrathmell, S. 2007. '"It's More Than Just What it is": Defetishising Commodities, Expanding Fields, Mobilising Change…' *Geoforum*. 38, 1113–1126.

Copulsky, J. R. 2011. *Brand Resilience: Managing Risk and Recover in a High-Speed World*. New York: Palgrave MacMillan.

Crouch, C. 2007. 'Trade Unions and Local Development Networks.' *Transfer*. 13, 2, 211–224.

Cumbers, A. 2012. *Reclaiming Public Ownership: Making Space for Economic Democracy*. London: Zed Books.

Da Silva Lopes, T. 2002. 'Brands and the Evolution of Multinationals in Alcoholic Beverages.' *Business History*. 44, 3, 1–30.

Da Silva Lopes, T. and Duguid, P. 2010. *Trademarks, Brands and Competitiveness*. London: Routledge.

Danesi, M. 2006. *Brands*. Routledge: London.

David, P. A. 1994. 'Why are Institutions the 'Carriers of History'? Path Dependence and the Evolution of Conventions, Organisations and Institutions.' *Structural Change and Economic Dynamics*. 5, 2, 205–220.

David, R. 2007. 'What Visibility for Wales? Connecting with the Consumer.' Memorandum from Institute of Welsh Affairs submitted to the Welsh Affairs Select Committee Inquiry *Globalisation and its Impact on Wales* (GLOB 12). Accessed 26 November 2014. http://www.publications.parliament.uk/pa/cm200607/cmselect/cmwelaf/ucglobal/m8.htm.

Dawley, S. 2014. 'Creating New Paths? Development of Offshore Wind, Policy Activism and Peripheral Region Development.' *Economic Geography*. 90, 1, 91–112.

de Chernatony, L. 2010. *From Brand Vision to Brand Evaluation* (2nd Edition). Amsterdam: Elsevier.

de Chernatony, L. and Dall'Olmo Riley, F. 1998. 'Modelling the Components of the Brand.' *European Journal of Marketing*. 32, 11/12, 1074–1090.

de Chernatony, L. and McDonald, M. 1998. *Creating Powerful Brands in Consumer, Service and Industrial Markets*. Oxford: Butterworth Heinemann.

Dedrick, J., Kraemer, L. and Linden, G. 2009. 'Who Profits from Innovation in Global Value Chains? A Study of the iPod and Notebook PCs.' *Industrial and Corporate Change*. 19, 1, 81–116.

Department for Environment, Food and Rural Affairs (DEFRA). 2006. National Application No: 02621 – Newcastle Brown Ale. London: DEFRA.

Dicken, P. 2011. *Global Shift* (6th edition). New York: Guilford Press.

Dickson, M. 2005. 'Rose Marie's Baby, from Geek to Chic.' *The Financial Times*. 5 October, London.

Dobson, S. and Merrington, J. 1977. *The Little Broon Book*. Gosforth: Geordieland Press.

Dossani, R. and Kenney, M. 2006. 'Software Engineering: Globalization and Its Implications, Paper for National Academy of Engineering "Workshop on the Offshoring of Engineering: Facts, Myths, Unknowns and Implications", 24–25 October, Washington DC.' Accessed 26 November 2014. http://www.nae.edu/File.aspx?id=10281.

Dossani, R. and Kenney, M. 2007. 'The Next Wave of Globalization: Relocating Service Provision to India.' *World Development*. 35, 5, 772–791.

Du Gay, P. and Pryke, M. 2002. *Cultural Economy: Cultural Analysis and Commercial Life*. London: Sage.

Duhigg, C. and Bradsher, K. 2012. 'How the U.S. Lost Out on iPhone Work.' *The New York Times*, 21 January, New York.

Dulleck, U., Kerschbamer, R. and Sutter, M. 2010. 'The Economics of Credence Goods: An Experiment on the Role of Liability, Verifiability, Reputation and Competition." Unpublished Paper.' Accessed 26 November 2014. http://ibe.eller.arizona.edu/docs/2010/Dulleck/AER_20090648_Manuscript.pdf.

Dwyer C and Jackson, P. 2003. 'Commodifying Difference: Selling EASTern Fashion.' *Environment and Planning D*. 21: 269–291.

Economic Geography. 2011. 'Emerging Themes in Economic Geography: Outcomes of the Economic Geography 2010 Workshop.' *Economic Geography*. 87, 2, 111–126.

Edensor, T. and Kothari, U. 2006. 'Extending Networks and Mediating Brands: Stallholder Strategies in a Mauritian Market.' *Transactions of the Institute of British Geographers*. 31, 323–336.

Elgan, M. 2012. 'Why Does Apple Inspire So Much Hate?' Blog post, 9 June. Accessed 26 November 2014. http://www.cultofmac.com/172428/why-does-apple-inspire-so-much-hate.

Ermann, U. 2011. 'Consumer Capitalism and Brand Fetishism: The Case of Fashion Brands in Bulgaria.' In *Brands and Branding Geographies*, edited by A. Pike, 107–125. Cheltenham: Elgar.

European Commission. 2006. 'Article 17(2) "Newcastle Brown Ale" EC No: UK/017/0372/1608.2004.' *Official Journal of the European Union C*. 280/13, 18 November 2006.

European Commission. 2013. 'Taxation and Customs Union.' Accessed 26 November 2014. http://ec.europa.eu/taxation_customs/customs/customs_duties/rules_origin/introduction/index_en.htm.

EuroStat. 2008. European Statistics Database. Accessed 28 November 2014. http://epp.eurostat.ec.europa.eu/portal/page/portal/eurostat/home/.

Fanselow, F. S. 1990. 'The Bazaar Economy or How Bizarre is the Bazaar Really?' *Man*. 25, 2, 250–265.

Faulconbridge, J., Beaverstock, J. V., Nativel, C. and Taylor, P. J. 2011. *The Globalization of Advertising: Agencies, Cities and Spaces of Creativity*. Abingdon: Routledge.

Finch, J. and May, T. 1998. 'Reputations: Putting a Zip in a Burberry.' *The Guardian*, 27 June, London.

Fleming, D. K. and Roth, R. 1991. 'Place in Advertising.' *The Geographical Review*. 81, 3, 281–291.

Frank, R. H. 2000. *Luxury Fever: Why Money Fails to Satisfy in an Era of Excess*. Princeton, NJ: Princeton University Press.

Frank, T. 1998. *The Conquest of Cool: Business Culture, Counterculture and the Rise of Hip Consumerism*. University of Chicago Press, IL: Chicago.

Friedman, T. 2005. *The World is Flat: A Brief History of the Twenty First Century*. New York: Farrar, Strauss and Giroux.

Froggatt, T. 2004. 'Building Brand Equity, Chief Executive's Presentation to ABN Amro Conference, 27 April.

Froud, J., Johal, S., Leaver, A. and Williams, K. 2012. 'Apple Business Model: Financialization Across the Pacific.' *CRESC Working Paper*, No. 111, April. Manchester: CRESC, University of Manchester.

Fuchs, D. 2006. 'Spice is right as La Mancha Relaunches Saffron as Luxury Brand.' *The Guardian*, 14 November, London.

Gereffi, G., Humphrey, J. and Sturgeon, T. 2005. 'The Governance of Global Value Chains.' *Review of International Political Economy*. 12, 1, 78–104.

Gibbon, P. and Ponte, S. 2008. 'Global Value Chains: From Governance to Governmentality?' *Economy and Society*. 37, 3, 365–392.

Glasmeier, A. 2000. *Manufacturing Time: Global Competition in the World Watch Industry, 1795–2000*. New York: Guilford Press.

Godsell, M. 2007. 'Is Royal Patronage Still Relevant?' 17 January, *Marketing*. London: Haymarket.

Goldman, R. and Papson, S. 1998. *Nike Culture: The Sign of the Swoosh*. Thousand Oaks, CA: Sage.

Goldman, R. and Papson, S. 2006. 'Capital's Brandscapes.' *Journal of Consumer Culture*. 6, 3, 327–353.

Goodrum, A. 2005. *The National Fabric: Fashion, Britishness, Globalization*. Oxford: Berg.

Gough, P. 2012. 'Banksy: The Bristol legacy.' In *Banksy: The Bristol Legacy*, edited by P. Gough. Bristol: Sansom and Company.

Go, F. and Govers, R. 2010. *International Place Branding Yearbook*. New York: Palgrave MacMillan.

Grabher, G. 1993. 'The Weakness of Strong Ties: The Lock-in of Regional Development in the Ruhr Area.' In *The Embedded Firm On the Socio-Economics of Interfirm Relations*, edited by G. Grabher, 255–278. London: Routledge.

Grabher, G. 2001. 'Ecologies of Creativity: The Village, The Group and the Heterarchic Organisation of the British Advertising Industry.' *Environment and Planning A*. 33, 351–374.

Grabher, G. 2006. 'Trading Routes, Bypasses, and Risky Intersections: Mapping the Travels of "Networks" Between Economic Sociology and Economic Geography.' *Progress in Human Geography*. 30, 2, 1–27.

Greenberg, M. 2008. *Branding New York: How a City in Crisis Was Sold to the World*. New York: Routledge.

Greenberg, M. 2010. 'Luxury and Diversity: Re-branding New York in the Age of Bloomberg.' In *Blowing up the Brand: Critical Perspectives on Promotional Culture*, edited by M. Aronczyk and D. Powers. Peter Lang: New York.

Griffiths, M. 2004. *Guinness is Guinness: The Colourful Story of a Black and White Brand*. London: Cyan Books.

Gross, D. 2006. 'To Chav and Chav not: Can Burberry Save Itself from the Tacky British Yobs Who Love It?' *The Slate*, July, Washington DC.

Gumbel, P. 2007. 'Burberry's New Boss Doesn't Wear Plaid.' *Fortune*. 156, 8, 124–130.

Hackney Today. 2006. 'Hackney 1, Nike 0.' *Hackney Today*. 142, 11 September.

Hadjimichalis, C. 2006. 'The End of the Third Italy as We Knew It.' *Antipode*. 38, 1, 82–106.

Haig, M. 2004a. *Brand Royalty: How the World's Top 100 Brands Thrive and Survive*. London: Kogan Page.

Haig, M. 2004b. *Brand Failures: The Truth Behind the 100 Biggest Branding Mistakes of All Time*. London: Kogan Page.

Halkier, A. and Therkelsen, H. 2011. 'Branding Provicail Cities: The Politics of Inclusion Strategy and Commitment.' In *Brands and Branding Geographies*, edited by A. Pike, 200–213. Cheltenham: Elgar

Han, M. C. 1989. 'Country Image: Halo or Summary Construct?' *Journal of Marketing Research*. XXVI, May, 222–229.

Hankinson, G. 2004. 'The Brand Images of Tourism Destinations: A Study of the Saliency of Organic Images.' *Journal of Product and Brand Management*. 1, 6–14.

Hannigan, J. 2004. 'Boom Towns and Cool Cities: The Perils and Prospects of Developing a Distinctive Urban Brand in a Global Economy.' Unpublished Paper from Leverhulme International Symposium: The Resurgent City, 19–21 April, LSE, London.

Harding, R. and Paterson, W. E. 2000. *The Future of the German Economy*. Manchester: Manchester University Press.

Hart, S. and Murphy, J. 1998. *Brands*. Basingstoke: Macmillan.

Hartwick, E. 2000. 'Towards a Geographical Politics of Consumption.' *Environment and Planning A*. 32, 1177–1192.

Harvey, D. 1989. *The Condition of Postmodernity*. Oxford: Blackwell.

Harvey, D. 1990. 'Between Space and Time: Reflections on the Geographical Imagination.' *Annals of the Association of American Geographers*. 80, 3, 418–434.

Harvey, D. 1996. *Justice, Nature and the Geography of Difference*. Oxford: Blackwell.

Harvey, D. 2002. 'The Art of Rent: Globalization, Monopoly and the Commodification of Culture.' In *A World of Contradictions: Socialist Register*, edited by L. Panitch and C. Leys, 93–110. London: Merlin Press.

Harvey, D. 2006. 'Neo-liberalism as Creative Destruction.' *Geografiska Annaler*. 88 B, 2, 145–158.

Hauge, A. 2011. 'Sports Equipment; Mixing Performance with Brands – the Role of the Consumers.' In *Brands and Branding Geographies*, edited by A. Pike, 91–106. Cheltenham: Elgar.

Hauge, A., Malmberg, A. and Power, D. 2009. 'The Spaces and Places of Swedish Fashion.' *European Planning Studies* 17, 4, 529–547.

Hayes, D. J., Lence, S. H. and Stoppa, A. 2004. 'Farmer-Owned Brands?' *Agribusiness.* 20, 3, 269–285.

Healy, A. 2012. 'Apple in Cork: Timeline.' *Evening Echo*, 20 April. Accessed 26 November 2014. http://www.eveningecho.ie/2012/04/20/apple-in-cork-timeline/.

Hebdige, D. 1989. *Hiding in the Light: One Images and Things*. London: Routledge.

Heineken. 2012. 'Annual Report.' Accessed 26 November 2014. www.annualreport.heineken.com.

Henderson, J., Dicken, P., Hess, M., Coe, N. and Yeung, H. W. C. 2002. 'Global Production Networks and the Analysis of Economic Development.' *Review of International Political Economy.* 9, 436–464.

Hill, A. 2007. 'Burberry's Reality Check.' 25 May, *The Financial Times*, London.

Hille, K. 2010. 'Foxconn to Move Some of its Apple Production.' *The Financial Times*, 29 June, London.

Hille, K. 2013. 'Huawei Looks to Dial a Different Number.' *The Financial Times*, 29 April, London.

Hille, K. and Jacob, R. 2013. 'Foxconn Plans Chinese Union Vote.' *The Financial Times*, 3 February, London.

Hilton, S. 2003. 'The Social Value of Brands.' In *Brands and Branding*, edited by R. Clifton and J. Simmons, 47–64. London: The Economist Ltd.

Hoad, P. 2012. 'What Next for the Global Blockbuster?' *The Guardian*, 26 July, London.

Hodgson, T. G. 2005. *Tyne Brewery: A Pictorial History, 1884–2005*. Edinburgh: Newcastle upon Tyne, Scottish and Newcastle plc.

Hollands, R. and Chatterton, P. 2003. 'Producing Nightlife in the New Urban Entertainment Economy.' *International Journal of Urban and Regional Research.* 27, 2, 361–385.

Hollis, N. 2010. *The Global Brand: How to Create and Develop Lasting Brand Value in the World Market*. New York: Palgrave MacMillan.

Holt, D. 2004. *How Brands Become Icons: The Principles of Cultural Branding*. Boston, MA: Harvard Business School Press.

Holt, D. 2006a. 'Toward a Sociology of Branding.' *Journal of Consumer Culture.* 6, 3, 299–302.

Holt, D. 2006b. 'Jack Daniel's America: Iconic Brands as Ideological Parasites and Proselytizers.' *Journal of Consumer Culture.* 6, 3, 355–377.

Holt, D. B., Quelch, J. A. and Taylor, E. L. 2004. 'How Global Brands Compete.' *Harvard Business Review*, September, 68–75.

Hudson, R. 1989. *Wrecking a Region: State Policies, Party Politics and Regional Change in North East England*. London: Pion.

Hudson, R. 2005. *Economic Geographies*. London: Sage.

Hudson, R. 2008. 'Cultural Political Economy Meets Global Production Networks: A Productive Meeting?' *Journal of Economic Geography.* 8, 421–440.

Hughes, A. 2006. 'Geographies of Exchange and Circulation: Transnational Trade and Governance.' *Progress in Human Geography.* 30, 5, 635–643.

Hughes, A. and Reimer, S. 2004. 'Introduction.' In *Geographies of Commodity Chains*, edited by A. Hughes and S. Reimer, 1–16. London: Routledge.

Huish, R. 2006. 'Logos a Thing of the Past? Not So Fast, World Social Forum!' *Antipode.* 38, 1, 1–6.

Humphrey, J. 2004. 'Upgrading in Global Value Chains.' Working Paper No. 28, Policy Integration Department, World Commission on the Social Dimension of Globalization, International Labour Office: Geneva. Accessed 26 November 2014. http://www.ilo.int/wcmsp5/groups/public/---dgreports/---integration/documents/publication/wcms_079105.pdf.

Hunter, J. 2001. *The Islanders and the Orb: The History of the Harris Tweed Industry, 1835–1995.* Stornoway: Acair.

Ibeh, K. I. N., Luo, Y., Dinnie, K. and Han, M. 2005. 'E-branding Strategies of Internet Companies: Some Preliminary Insights from the UK.' *Journal of Brand Management.* 12, 5, 355–373.

Ilbery, B. and Kneafsey, M. 1999. 'Niche Markets and Regional Speciality Food Products in Europe: Towards a Research Agenda.' *Environment and Planning* A. 31, 2207–2222.

Interbrand. 2010. *Best Global Brands 2010.* London: Interbrand.

Interbrand. 2012. *Best Global Brands 2012.* London: Interbrand.

Isaacson, W. 2011. *Steve Jobs.* London: Little, Brown.

iSuppli. 2013. 'Preliminary Bill of Matierials (BOM) Estimate for the 16GB Version of the iPhone 4.' Accessed 26 November 2014. http://www.isuppli.com/Teardowns/News/Pages/iPhone-4-Carries-Bill-of-Materials-of-187-51-According-to-iSuppli.aspx.

Jackson, P. 1999. 'Commodity Cultures: The Traffic in Things.' *Transactions of the Institute of British Geographers.* 24, 95–108.

Jackson, P. 2002. 'Commercial Cultures: Transcending The Cultural and the Economic.' *Progress in Human Geography.* 26, 3–18.

Jackson, P. 2004. 'Local Consumption Cultures in a Globalizing World.' *Transactions of the Institute of British Geographers.* 29, 165–178.

Jackson, P., Russell, P. and Ward, N. 2007. 'The Appropriation of "Alternative" Discourses by "Mainstream" Food Retailers.' In *Alternative Food Geographies: Representation and Practice,* edited by D. Maye, L. Holloway and M. Kneafsey, 309–330. Amsterdam: Elsevier.

Jackson, P., Russell, P. and Ward, N. 2011. 'Brands in the Making: A life History Approach.' In *Brands and Branding Geographies,* edited by A. Pike, 59–74. Cheltenham: Elgar.

Jessop, B. 2008. 'Discussant Comments on Adam Arvidsson's Paper "Brand and General Intellect", ESRC "Changing Cultures of Competitiveness"' Seminar Series, Institute of Advanced Studies, Lancaster University.

Johansson, J. K. 1993. 'Missing a Strategic Opportunity: Manager's Denial of Country-of-Origin Effects.' In *Product Country Images: Impact and Role in International Marketing,* edited by N. Papadopoulos and L. A. Heslop, 77–86. New York: International Business Press.

Jones, R. M. and Hayes, S. G. 2004. 'The UK Clothing Industry: Extinction or Evolution?' *Journal of Fashion Marketing and Management.* 8, 3, 262–278.

Julier, G. 2005. 'Urban Designscapes and the Production of Aesthetic Consent.' *Urban Studies.* 42, 5/6, 869–887.

Just Drinks. 2006. 'Newcastle Brown Ale Now #1 imported ale.'

Kahney, L. 2002. 'Apple: It's all About the Brand.' *Wired,* 12 April. Accessed 26 November 2014. http://www.wired.com/gadgets/mac/commentary/cultofmac/2002/12/56677.

Kahney, L. 2005. *The Cult of iPod.* San Francisco, CA: No Starch Press.

Kahney, L. 2006. *The Cult of Mac.* San Francisco, CA: No Starch Press.

Kapferer, J.-N. 2002. 'Is There Really No Hope for Local Brands?' *Brand Management.* 9, 3, 163–170.

Kapferer, J.-N. 2005. 'The Post-Global Brand.' *Brand Management.* 12, 5, 319–324.

Karlgaard, R. 2012. 'Kodak Didn't Kill Rochester. It Was the Other Way Round.' *The Wall Street Journal,* 14 January.

Keller, K. 2003. 'Brand Synthesis: The Multidimensionality of Brand Knowledge.' *Journal of Consumer Research.* 29, 4, 595–600.

Kemeny, T. and Rigby, D. 2012. 'Trading Away What Kind of Jobs? Globalization, Trade and Tasks in the US Economy.' *Review of World Economics.* 148, 1, 1–16.

Klein, N. 2000. *No Logo.* London: Flamingo.

Klein, N. 2010. *No Logo* (2nd edition). Fourth Estate: London.

Klein, B. and Leffler K. 1981. 'The Role of Market Forces in Assuring Contractual Performance.' *Journal of Political Economy.* 89, 615–641.

Klingman, A. 2007. *Brandscapes: Architecture in the Experience Economy*. Cambridge, MA: MIT Press.

Koehn, N. 2001. *Brand New: How Entrepreneurs Earned Consumers' Trust from Wedgwood to Dell.* Boston, MA: Harvard Business School Press.

Kopytoff, I. 1986. 'The Cultural Biography of Things: Commoditization as Process.' In *The Social Life of Things*, edited by A. Apparadurai, 64–91. Cambridge: Cambridge University Press.

Kornberger, M. 2010. *Brand Society: How Brands Transform Management and Lifestyle.* Cambridge: Cambridge University Press.

Krishna, K. 2005. 'Understanding Rules of Origin.' NBER Working Paper No. 11149, National Bureau of Economic Research. Accessed 26 November 2014. www.nber.org/papers/w11150.pdf.

Kwong, R. 2008. 'Ateliers of Asia Stake Their Claim.' 29 May, *The Financial Times*, London.

Lash, S. and Urry, J. 1994. *Economies of Signs and Space*. London: Sage.

Lashinsky, A. 2012. *Inside Apple: The Secrets Behind the Past and Future Success of Steve Jobs's Iconic Brand.* London: John Murray.

Lawson, N. 2006. 'Turbo-Consumerism is the Driving Force Behind Crime.' *The Guardian*, 29 June, London.

Lazonick, W. 2010. 'Innovative Business Models and Varieties of Capitalism: Financialization of the U.S. Corporation.' *Business History Review*. 84, 675–702.

Lee, R. 2002. 'Nice Maps, Shame About the Theory"? Thinking Geographically About the Economic.' *Progress in Human Geography.* 26, 3, 333–355.

Lee, R. 2006. 'The Ordinary Economy: Tangled Up in Values and Geography.' *Transactions of the Institute of British Geographers NS.* 31, 413–432.

Levitt, T. 1983. 'The Globalization of Markets.' *Harvard Business Review*. May–June, 92–102.

Lewis, N. 2007. 'Micro-practices of Globalizing Education: Branding.' Paper for the Second Global Conference on Economic Geography, 25–28 June, Beijing, China.

Lewis, N., Larner, W. and Le Heron, R. 2008. 'The New Zealand Designer Fashion Industry: Making Industries and Co-constituting Political Projects.' *Transactions of the Institute of British Geographers NS.* 33, 42–59.

Lindemann, J. 2010. *The Economy of Brands*. Basingstoke: Palgrave Macmillan

Littler, C. 2005. 'Beyond the Boycott: Anti-consumerism, Cultural Change and the Limits of Reflexivity.' *Cultural Studies.* 19, 2, 227–252.

Long, X. 2012. 'Designs on the Best of British.' *The Financial Times*, 23 February, London.

Lury, C. 2004. *Brands: The Logos of the Global Economy*. London: Routledge.

Lury C. 2011. 'Brands: Boundary Method Objects and Media Space.' In *Brands and Branding Geographies*, edited by A. Pike, 44–59. Cheltenham: Elgar.

Lury, C. and Moor, L. 2010. 'Brand Valuation and Topological Culture.' In *Blowing Up the Brand: Critical Perspectives on Promotional*, edited by M. Aronczyk and D. Powers. New York: Peter Lang.

McCracken, G. 1993. 'The Value of the Brand: An Anthropological Perspective.' In *Brand Equity and Advertising*, edited by D. A. Aaker and A. L. Biel. Hillsdale, NJ: Lawrence Erlbaum Associates.

McDermott, C. 2002. *Made in Britain: Tradition and Style in Contemporary British Fashion*, London: Mitchell Beazley.

McFall, L. 2002. 'Advertising, Persuasion and the Culture/Economy Dualism.' In *Cultural Economy: Cultural Analysis and Commercial*, edited by P. Du Gay and M. Pryke, 148-165. London: Life Sage.

MacKinnon, D., Cumbers, A., Pike, A., Birch, K. and McMaster, R. 2009. 'Evolution in Economic Geography.' *Economic Geography.* 85, 2, 175–182.

McLaren, R., Tyler, D. J. and Jones, R. M. 2002. 'Parade – Exploiting the Strengths of 'Made in Britain' Supply Chain.' *Journal of Fashion Marketing and Management.* 6, 1, 35–43.

McLaughlin, K. 2010. 'Apple COO: We're A Mobile Device Company. CRN News.' Accessed26 November 2014. http://www.crn.com/news/mobility/223100456/apple-coo-were-a-mobile-device-company.htm.

McRobbie, A. 1998. *British Fashion Design: Rag Trade or Image Industry?* London: Routledge.

Mair, A., Florida, R. and Kenney, M. 1988. 'The New Geography of Automobile Production: Japanese Transplants in North America.' *Economic Geography.* 63, 4, 353–373.

Marketing Minds. 2013. Apple's Branding Strategy.

Markusen, A. 2007. *Reining in the Competition for Capital.* Kalamazoo, MI: W. E. Upjohn Institute for Employment Research.

Marx, K. 1976. *Capital Volume I* (trans. B. Fowkes). Harmondsworth: Penguin.

Maurer, A. 2013. 'Trade in Value Added: What is the Country of Origin in an Interconnected World?' Accessed 26 November 2014. http://www.wto.org/english/res_e/statis_e/miwi_e/background_paper_e.htm.

Menkes, S. 2010. 'Throwing Down the Gauntlet.' *New York Times*, 28 September. Accessed 26 November 2014. http://www.nytimes.com/2010/09/29/fashion/29iht-rprada.html.

Middlebrook, S. 1968. *Newcastle upon Tyne: Its Growth and Achievement* (2nd edition). Wakefield: S. R. Publishers Ltd.

Miller, D. 1998. 'Coca-cola: A Black Sweet Drink from Trinidad.' In *Material Culture*, edited by D. Miller, 169–187. London: Routledge.

Miller, D. 2002. 'Turning Callon the Right Way Up.' *Economy and Society.* 31, 2, 218–233.

Miller, D. 2008. 'Reply to Wengrow "Prehistories of Commodity Branding".' *Current Anthropology.* 49, 1, 23.

Milne, R. 2008. 'High Quality Can Beat the Credit Crisis.' 29 May, *The Financial Times*, London.

Molotch, H. 2002. 'Place in Product.' *International Journal of Urban and Regional Research.* 26, 4, 665–688.

Molotch, H. 2005. *Where Stuff Comes From: How Toasters, Toilets, Cars, Computers and Many Other Things Come to Be As They Are.* New York: Routledge.

Montague, D. 2002. 'Stolen Goods: Coltan and Conflict in the Democratic Republic of Congo.' *SAIS Review.* XXII, 1, 1–16.

Moor, L. 2007. *The Rise of Brands.* London: Berg.

Moor, L. 2008. 'Branding Consultants as Cultural Intermediaries.' *The Sociological Review.* 56, 3, 408–428.

Moore, C. M. and Birtwistle, G. 2004. 'The Burberry Business Model: Creating an International Luxury Fashion Brand.' *International Journal of Retail and Distribution Management.* 32, 8, 412–422.

Moran, W. 1993. 'The Wine Appellation as Territory in France and California.' *Annals of the Association of American Geographers.* 82, 3, 27–49.

Morello, G. 1984. 'The 'Made-In' issue – A Comparative Research on the Image of Domestic and Foreign Products.' *European Research.* July, 95–100.

Morello, G. 1993. 'International Product Competitiveness and the "Made in" Concept.' In *Product Country Images: Impact and Role in International Marketing*, edited by N. Papadopoulos and L. A. Heslop, 285–309. New York: International Business Press.

Morgan, K., Marsden, T. and Murdoch, J. 2006. *Worlds of Food.* Oxford: Oxford University Press.

Moritz, M. 2009. *Return to the Little Kingdom: Steve Jobs, the Creation of Apple, and How it Changed the World* (2nd Edition). London: Duckworth Overlook.

Mudambi, R. 2008. 'Location, Control and Innovation in Knowledge-Intensive Industries.' *Journal of Economic Geography.* 8, 5, 699–725.
Muñiz, A. and O'Guinn, T. 2001. 'Brand Community.' *Journal of Consumer Research.* 27, 4, 412–432.
Murphy, J. 1998. 'What is Branding?' In *Brands*, edited by S. Hart and J. Murphy, 1–12. Basingstoke: Macmillan.
Murray, J. 1998. 'Branding in the European Union.' In *Brands*, edited by S. Hart and J. Murphy, 135–151. Basingstoke: Macmillan.
Nakamura, L. 2003. 'A Trillion Dollars a Year in Intangible Investment and the New Economy.' In *Intangible Assets: Values, Measures and Risks*, edited by J. R. M. Hand and B. Lev. New York: Oxford University Press.
Nayak, A. 2003. 'Last of the 'Real Geordies'?: White Masculinities and the Sub-Cultural Response to De-industrialisation.' *Environment and Planning D: Society and Space.* 21, 1, 7–25.
NESTA. 2011. *Driving Economic Growth: Innovation, Knowledge Spending and Productivity Growth in the UK.* London: NESTA.
Newcastle Brown Ale. 2007. 'The History of Newcastle Brown Ale.' Accessed 26 November 2014. www.newcastlebrownale.co.uk.
Neilsen, J. and Pritchard, B. 2011. *Value Chain Struggles* .Chichester: John Wiley and Sons.
Noble, S. 2011. 'Marketers – Purveyors of Puffery or the Engines of New Growth?' *The Guardian*, 14 November, London.
Norris, D. G. 1993. '"Intel Inside" Branding a Component in a Business Market.' *Journal of Business and Industrial Marketing.* 8, 1, 14–24.
NPR Staff. 2012. 'Made in the USA: Saving the American Brand.' 28 January. Accessed 26 November 2014. http://www.npr.org/2012/01/28/146033135/made-in-the-usa-saving-the-american-brand
Nussbaum, B. 2009. 'iPhones in China Don't Say they are Assembled in China.' Bloomberg Business Week, 30 November. Accessed 21 September 2010. http://www.businessweek.com/innovate/NussbaumOnDesign/archives/2009/11/iphones_in_china_dont_say_they_are_assembled_in_china.html.
O'Neill, P. M. 2011. 'The Language of Local and Regional Development.' In *Handbook of Local and Regional Development*, edited by A. Pike, A. Rodríguez-Pose and J. Tomaney. London: Routledge.
Ohmae, K. 1992. *The Borderless World.* New York: McKinsey and Company.
Okonkwo, O. 2007. *Luxury Fashion Branding: Trends, Tactics, Techniques.* New York: Palgrave Macmillan.
Olins, W. 2003. *On Brand.* New York: Thames and Hudson.
Osborne, G. 2011. 'Budget Statement 2011.'Accessed 26 November 2014. http://www.parliament.uk/business/news/2011/march/budget-2011-statement/.
Packard V. 1980. *The Hidden Persuaders* (2nd edition). Brooklyn, NY: Pocket Books.
Pallota, D. 2011. 'A Logo is Not a Brand.' HBR Blog Network, 15 June. Accessed 26 November 2014. http://blogs.hbr.org/pallotta/2011/06/a-logo-is-not-a-brand.html.
Papadopoulos, N. 1993. 'What Product and Country Images Are and Are Not.' In *Product Country Images: Impact and Role in International Marketing*, edited by N. Papadopoulos and L. A. Heslop, 3–38. New York: International Business Press.
Papadopoulos, N. and Heslop, L. A. 1993. *Product Country Images: Impact and Role in International Marketing.* New York: International Business Press.
Parrott, N., Wilson, N. and Murdoch, J. 2002. 'Spatializing Quality: Regional Protection and the Alternative Geography of Food.' *European Urban and Regional Studies.* 9, 3, 241–261.
Pasquinelli, C. (2014) 'Branding as Urban Collective Strategy-Making: The Formation of NewcastleGateshead's Organisational Identity', *Urban Studies.* 51, 4, 727–743.

Pearson, G. 1999. *Sex, Brown Ale and Rhythm and Blues.* Newcastle: Snaga Publications.
Peck, J. 2012. 'Economic Geography: Island Life.' *Dialogues in Human Geography.* 2, 2, 113–133.
Peck, J. and Theodore, N. 2007. Variegated Capitalism. *Progress Human Geography.* 31, 6, 731–772.
Perrons, D. 1999. 'Reintegrating Production and Consumption, Or Why Political Economy Still Matters.' In *Critical Development Theory: Contributions to New Paradigm*, edited by R. Munck and D. O'Hearn, 91–112. London and New York: Zed Books.
Peters, T. 1999. *The Brand You 50.* New York: Random House.
Phau, I. and Prendergast, G. 1999. 'Tracing the Evolution of Country of Origin Research: In Search of New Frontiers.' *Journal of International Marketing and Exporting.* 4, 2, 71–83.
Phau, I. and Prendergast, G. 2000. 'Conceptualizing the Country of Origin of Brand.' *Journal of Marketing Communications.* 6, 159–170.
Pickles, J. and Smith, A. 2011. 'De-localisation and Persistence in the European Clothing Industry: The Reconfiguration of Production Networks.' *Regional Studies.* 45, 2, 167–185
Pike, A. 2007. 'Editorial: Whither Regional Studies?' *Regional Studies.* 41, 9, 1143–1148.
Pike, A. 2009a. 'Geographies of Brands and Branding.' *Progress in Human Geography.* 33, 619–645.
Pike, A. 2009b. 'Brand and Branding Geographies.' *Geography Compass.* 3, 1, 190–213.
Pike, A. 2009c. 'De-industrialization.' In *International Encyclopedia of Human Geography*, edited by R. Kitchin and N. Thrift, 51–59. Amsterdam: Elsevier.
Pike, A. 2010. *Origination*, Inaugural Lecture, Wednesday 10 November 2010, Great North Museum, Newcastle upon Tyne, UK.
Pike, A. 2011a. 'Placing Brands and Branding: A Socio-Spatial Biography of Newcastle Brown Ale.' *Transactions of the Institute of British Geographers.* 36, 206–222.
Pike, A. 2011b. *Brands and Branding Geographies.* Cheltenham: Elgar.
Pike, A. 2011c. 'Introduction: Brand and Branding Geographies.' In *Brands and Branding*, edited by A. Pike, 3–24. Cheltenham: Elgar.
Pike, A. 2011d. 'Conclusions: Brand and Branding Geographies.' In *Brands and Branding*, edited by A. Pike, 324–337. Cheltenham: Elgar.
Pike, A. 2013. 'Economic Geographies of Brands and Branding.' *Economic Geography.* 89, 4, 317–339.
Pike, A. and Pollard, J. 2010. 'Economic Geographies of Financialization.' *Economic Geography.* 86, 1, 29–51.
Pike, A., Dawley, S. and Tomaney, J. 2010. 'Resilience, adaptation and adaptability.' *Cambridge Journal of Regions, Economy and Society.* 3, 1, 59–70
Pike, A., Rodríguez-Pose, A. and Tomaney, J. 2006. *Local and Regional Development.* London: Routledge.
Pike, A. Rodríguez-Pose, A. and Tomaney, J. 2007. 'What kind of local and regional development and for whom?' *Regional Studies.* 41, 9, 1253–1269.
Pike, A., Rodríguez-Pose, A. and Tomaney, J. 2011. *Handbook of Local and Regional Development*, London: Routledge.
Pine, B. J. and Gilmore, J. H. 1999. *The Experience Economy Work is Theatre and Every Business a Stage.* Boston, MA: Harvard Business School Press.
Power, D. and Hauge, A. 2008. 'No Man's Brand – Brands, Institutions, Fashion and the Economy.' *Growth and Change.* 39, 1, 123–143.
Power, D. and Jansson, J. 2011. 'Constructing Brands form the Outside? Brand Channels Cyclical Clusters and Global Circuits.' In *Brands and Branding Geographies*, edited by A. Pike, 125–150. Cheltenham: Elgar.
Prince, M. and Plank, W. 2012. 'A Short History of Apple's Manufacturing in the U.S.' *The Wall Street Journal*, 6 December. New York: Dow Jones and Company Inc.

Quelch, J. and Jocz, K. 2012. *All Business is Local: Why Place Matters More Than Ever in a Global, Virtual World*. New York: Penguin.
Ray, C. 1998. 'Culture, Intellectual Property and Territorial Rural Development.' *Sociologia Ruralis*. 38, 1, 3–20.
Reich, R. 1990. 'Who Is Us?' *Harvard Business Review*. 1, 1–11.
Reimer, S. and Leslie, D. 2008. 'Design, National Imaginaries and the Home Furnishings Commodity Chain.' *Growth and Change*. 39, 1, 144–171.
Relph, E. 1976. *Place and Placelessness*. London: Pion.
Ricca, M. 2008. 'The Luxury Kingdom.' In Interbrand, *Best Global Brands 2008*. London: Interbrand.
Richardson, R. 2012. Place Branding – Literature Review. Unpublished Paper. CURDS: Newcastle University.
Richardson, R., Belt, V. and Marshall, N. 2000. 'Taking Calls to Newcastle: The Regional Implications of the Growth in Call Centres.' *Regional Studies*, 34, 4, 357–369.
Ries, A. and Ries, L. 1998. *The 22 Immutable Laws of Branding*. London: HarperCollins Publishers.
Riezebos, R. 2003. *Brand Management*. Harlow: Pearson.
Rigby, R. 2012. 'Brands in the Social Lexicon.' 13 June, *The Financial Times*, London.
Ritson, M. 2008. 'Burberry Protest is no Brand Breaker.' *Marketing*. 21 March. London: Haymarket.
Ritzer, G. 1998. *The McDonaldization Thesis: Explorations and Extensions*. London: Sage.
Roberts, K. 2005. *Lovemarks: The Future Beyond Brands*. Brooklyn, NY: Powerhouse.
Room, A. 1998. 'History of Branding.' In *Brands*, edited by S. Hart and J. Murphy, 13–23. Basingstoke: MacMillan.
Ross, A. 2004. *Low Pay, High Profile: The Global Push for Fair Labor*. New York: New Press.
Roth, M. S. and Romeo, J. B. 1992. 'Matching Product Category and Country Image Perceptions: A Framework for Managing Country-of-Origin Effects.' *Journal of International Business Studies*. 23, 3, 477–497.
Ryssdal, K. 2009. 'Ads Seek to Rebrand "Made in China".' Accessed 26 November 2014. http://www.marketplace.org/topics/business/ads-seek-rebrand-made-china.
SABMiller. 2013. 'Annual Report.' SABMiller.
S&N. 2004. 'Reorganisation of Brewing Operations on Tyneside.' Press release, 22 April. Accessed 26 November 2014. www.scottish-newcastle.com/.
S&N. 2006. *Scottish and Newcastle plc Interim Report 2006*. Edinburgh: S&N.
S&N. 2007. 'Newcastle Brown Ale.' S&N.
Samsung. 2007. 'SAMSUNG Concludes Contract with the International Olympic Committee to Sponsor Olympic Games Through 2016 on Apr 23, 2007.' Accessed 26 November 2014. http://www.samsung.com/my/news/localnews/2007/samsung-concludes-contract-with-the-international-olympic-committee-to-sponsor-olympic-games-through-2016.
Sanger, D. E. 1984. 'New Plants May Not Mean New Jobs.' *The New York Times*, 25 March, New York.
Saxenian, A. 1996. *Regional Advantage: Culture and Competition in Silicon Valley and Route 128*. Cambridge, MA: Harvard University Press.
Saxenian, A. 1999. 'Comment on Kenney and von Burg, 'Technology, Entrepreneurship and Path Dependence: Industrial Clustering in Silicon Valley and Route 128.' *Industrial and Corporate Change*. 8, 1, 105–110.
Saxenian, A. 2005. 'From Brain Drain to Brain Circulation: Transnational Communities and Regional Upgrading in India and China.' *Studies in Comparative International Development*. 40, 2, 35–61.

Sayer, A. 1997. 'The Dialectic of Culture and Economy.' In *Geographies of Economies*, edited by R. Lee and J. Wills, 16–26. London: Arnold.
Sayer, A. 1999. 'Long Live Postdisciplinary Studies! Sociology and the Curse of Disciplinary Parochialism/Imperialism.' Department of Sociology Papers, Lancaster University.
Sayer, A. 2001. 'For a Critical Cultural Political Economy.' *Antipode*. 33, 4, 687–708.
Schiro, A.-M. 1999. 'Burberry Modernizes and Reinvents Itself.' *New York Times*, 5 January, New York.
Schroeder, J. and Salzer-Morling, M. (eds) 2001. *Brand Culture*. London: Routledge.
Silverstein, M. J. and Fiske, N. 2003. *Trading Up: The New American Luxury*. London: Portfolio.
Scott, A. J. 1998. *Regions and the World Economy*. Oxford: Oxford University Press.
Scott, A. J. 2000. *The Cultural Economy of Cities*. London: Sage.
Scott, A. J. 2007. 'Capitalism and Urbanization in a New Key? The Cognitive-Cultural Dimension.' *Social Forces*. 85, 4, 1465–1482.
Scott, A. J. 2010. 'Cultural Economy and the Creative Field of the City.' *Geografiska Annaler: Series B, Human Geography*. 92, 2, 115–130.
Sennett, R. 2006. *The Culture of the New Capitalism*. New Haven, CT: Yale University Press.
Sheth, J. 1998. 'Reflections of International Marketing: In Search of New Paradigms.' Keynote address of Marketing Exchange Colloquium, Vienna, Austria, 23–25 July.
Sheyder, E. 2012. 'Focus on Past Glory Kept Kodak from Digital Win.' *Reuters*, 19 January. Accessed 26 November 2014. http://www.reuters.com/article/2012/01/19/us-kodak-bankruptcy-idUSTRE80I1N020120119.
Sissons, A. 2011. 'More than Making Things: A New Future for Manufacturing in a Service Economy.' *Report for The Work Foundation*. London: The Work Foundation.
Slater, D. 2002. 'Capturing Markets from the Economists.' In *Cultural Economy: Cultural Analysis and Commercial Life*, edited by P. Du Gay and M. Pryke, 59–77. London: Sage.
Smith, A. and Bridge, G. 2003. 'Intimate Encounters: Culture-Economy-Commodity.' *Environment and Planning D: Society and Space*. 21, 257–268.
Smith, A., Rainnie, A., Dunford, M., Hardy, J., Hudson, R. and Sadler, D. 2002. 'Networks of Value, Commodities and Regions: Reworking Divisions of Labour in Macro-Regional Economies.' *Progress in Human Geography*. 26, 1, 41–63.
Smith, T. 2000. 'LG shuts down Welsh iMac Production Line.' *The Register*, 10 March. Accessed 26 November 2014. http://www.theregister.co.uk/2000/03/10/lg_shuts_down_welsh_imac.
Spence, S. 2002a. 'The Branding of Apple: Apple's Intangible Asset.' Accessed 26 November 2014. http://tidbits.com/article/6919.
Spence, S. 2002b. 'The Branding of Apple: The Retail Bridge.' Accessed 26 November 2014. http://tidbits.com/article/6926.
Spicer, A. 2010. 'Branded Life: A Review of Key Works on Branding.' *Organization Studies*. 31, 12, 1735–1740.
Storper, M. 1995. *The Regional World*. New York: Guilford Press.
Storper, M. and Venables, A. 2004. 'Buzz: Face-to-Face Contact and the Urban Economy.' *Journal of Economic Geography*. 4, 4, 351–370.
Streeck, W. 2012. 'Citizens as Consumers: Considerations on the New Politics of Consumption.' *New Left Review*. 76, 27–47.
Sum, N.-L. 2011. 'The Making and Recontextualizing of "Competitiveness" as a Knowledge Brand across Different Sites and Scales.' In *Brands and Branding Geographies*, edited by A. Pike, 165–184. Cheltenham: Elgar.
Sunley, P., Pinch, S., Reimer, S. and Macmillen, J. 2008. 'Innovation in a Creative Production System: The Case of Design.' *Journal of Economic Geography*. 8, 5, 675–698.
Taylor, P. 2004. *World City Network: A Global Urban Analysis*. London: Routledge.

Taylor, P., Mulvey, G., Hyman, J. and Bain, P. 2002. 'Work Organization, Control and the Experience of Work in Call Centres.' *Work, Employment and Society*. 16, 1, 133–150.

Thakor, M.V. and Kohli, C. S. 1996. 'Brand Origin: Conceptualization and Review.' *Journal of Consumer Marketing*. 13, 3, 27–42.

The Economic Times. 2012. 'Silicon Valley Plant Named as Apple Manufacturer.' *The Economic Times*, 26 January. Accessed 26 November 2014. http://economictimes.indiatimes.com/topic/Silicon-Valley-plant-named-as-Apple-manufacturer/.

The Economist. 2001. 'Stretching the Plaid.' 1 February, *The Economist*, London.

The Economist. 2005. '"Brand New World" in The World in 2005.' London.

The Economist. 2009. *Brands and Branding, Edited collection* (2nd edition). London: Profile Books.

Thode, S. F. and Maskulka, J. M. 1998. 'Place-Based Marketing Strategies, Brand Equity and Vineyard Valuation.' *Journal of Product and Brand Management*. 7, 5, 379–399.

Thomas, D. 2007. 'Made in Italy in the Sly.' *New York Times*, 23 November, New York.

Thomas, D. 2008. *How Luxury Lost its Lustre*. Harmondsworth: Penguin.

Thrift, N. 1996. *Spatial Formations*. London: Sage.

Thrift, N. 2005. *Knowing Capitalism*. London: Sage.

Thrift, N. 2006. 'Re-inventing Invention: New Tendencies in Capitalist Commodification.' *Economy and Society*. 35, 2, 279–306.

Tighe, C. 2004. 'Star Sees Sun Set on Northern Brewery.' *The Financial Times*, 23 April, London.

Tokatli, N. 2012a. 'Old Firms, New Tricks and the Quest for Profits: Burberry's Journey from Success to Failure and Back to Success Again.' *Journal of Economic Geography*. 12, 1, 55–77.

Tokatli, N. 2012b. 'The Changing Role of Place-Image in the Profit Making Strategies of the Designer Fashion Industries.' *Geography Compass*. 6, 1, 35–43.

Tokatli, N. 2013. 'Doing a Gucci: The Transformation of an Italian Fashion Firm into a Global Powerhouse in a "Los Angeles-izing" World.' *Journal of Economic Geography*. 13, 2, 239–255 (Special Issue: Global Retail and Global Finance – Honouring Neil Wrigley).

Tomaney, J. 2006. 'North East England: A Brief Economic History' Paper for the North East Regional Information Partnership (NERIP) Annual Conference, 6 September, Newcastle upon Tyne.

Tregear, A. 2003. 'From Stilton to Vimto: Using Food History to Re-think Typical Products in Rural Development.' *Sociologia Ruralis*. 43, 2, 91–107.

Tungate, M. 2005. *Fashion Brands: Branding Styles from Armani to Zara*. London and Sterling, VA: Kogan Page.

Turok, I. 2009. 'The Distinctive City: Pitfalls in the Pursuit of Differential Advantage.' *Environment and Planning A*. 41, 1, 13–30.

Turok, I. 2011. 'Inclusive growth: Meaningful Goal or Mirage?' In *Handbook of Local and Regional Development*, edited by A. Pike, A. Rodríguez-Pose and J. Tomaney. London: Routledge.

Upshaw, L. 1995. *Building Brand Identity: A Strategy for Success in a Hostile Marketplace*. New York: Wiley and Sons.

Urry, J. 1995. *Consuming Places*. London: Routledge.

Urry, J. 2003. *Global Complexity*. Cambridge: Polity Press.

van Ham, P. 2001. 'The Rise of the Brand State.' *Foreign Affairs*. 80, 5, 2–6.

van Ham, P. 2008. 'Place Branding: The State of the Art.' *The Annals of the American Academy of Political and Social Science*. 1–24.

Veblen, T. 1899. *The Theory of the Leisure Class: An Economic Study of Institutions*. New York: Macmillan.

Von Borries, F., Klincke, H., Polsa, B. and Museum fur Kunst und Gewerbe. 2011. *Apple Design: The History of Apple Design*. Ostfildern: Hatje Cantz.

WalesOnline. 2011. 'Burberry Workers Still Face Hardship Four Years On', 22 June. Accessed on 26 November 2014. http://www.walesonline.co.uk/news/local-news/burberry-workers-still-face-hardship-1828814.

Walker, D. and the Bay Area Study Group 1990. 'The Playground of US Capitalism? The Political Economy of the San Francisco Bay Area in the 1980s.' In *Fire in the Hearth: The Radical Politics of Place in America*, edited by M. Davis, S. Hiatt, M. Kennedy, S. Ruddick and M. Sprinker. London: Verso.

Walker, H. 2004a. 'Newcastle Brown kept on Tyne.' *The Journal*, 23 April, Newcastle upon Tyne.

Walker, H. 2004b. 'Boss Happy at Sober Response to Brown Move.' *The Journal*, 21 June, Newcastle upon Tyne.

Watts, D. C. H., Ilbery, B. and Maye, D. 2005. 'Making Reconnections in Agro-Food Geography: Alternative Systems of Food Provision.' *Progress in Human Geography*. 29, 1, 22–40.

Watts, J. 2002. 'The Once-Dowdy Brand is Determined to Maintain its Cachet.' *Campaign*. November, Haymarket: London.

Watts, M. 2005. 'Commodities.' In *Introducing Human Geographies* (2nd edition), edited by P. Cloke, P. Crang and M. Goodwin Hodder, 527–546. Abingdon: Arnold.

Weller, S. 2007. 'Fashion as Viscous Knowledge: Fashion's Role in Shaping Trans-National Garment Production.' *Journal of Economic Geography*. 7, 39–66.

Welsh Affairs Select Committee 2007. *Globalisation and its Impact on Wales*, Transcript of Oral Evidence, HC 281-iv, House of Commons: London.

Wengrow, D. 2008. 'Prehistories of Commodity Branding.' *Current Anthropology*. 49, 1, 7–34.

Western Mail. 2006. 'Burberry to close Welsh factory.' *Western Mail*, 6 September, Cardiff.

Whitfield, G. 2006. 'Dog is Slipping its Local Leash.' *The Journal*, 24 November, Newcastle upon Tyne

Whitten, N. 2007. 'Profits Soar But Jobs May Go.' *The Evening Chronicle*, 20 February, Newcastle upon Tyne.

Williams, R. 1980. *Problems in Materialism and Culture*. London: Verso.

Willmott, H. 2010. 'Creating "Value" Beyond the Point of Production: Branding, Financialization and Market Capitalization.' *Organization*. 17, 5, 517–542.

Woods, S. 2006. 'Google Top Gainer as Burberry Takes "Britishness" to the World.' *Brand Republic*, 28 July, London.

World Bank. 2012. 'Gross Domestic Product 2012.' Accessed 26 November 2014. http://databank.worldbank.org/data/download/GDP.pdf.

World Trade Organization. 2013. 'Rules of Origin.' Accessed 26 November 2014. http://www.wto.org/english/tratop_e/roi_e/roi_e.htm.

Wortzel, L. H. 1987. 'Retailing Strategies for Today's Mature Marketplace.' *Journal of Business Strategy*. 7, 4, 45–56.

Yeung, H. 2009. 'Regional Development and the Competitive Dynamics of Global Production Networks: An East Asian Perspective.' *Regional Studies*. 43, 3, 325–351.

Yoffie, D. B. and Rosanno, P. 2012. *Apple Inc. in 2012, HBS No. 9-712-490*. Boston, MA: Harvard Business School Publishing.

Index

Origination: The Geographies of Brands and Branding, First Edition. Andy Pike.
© 2015 John Wiley & Sons, Ltd. Published 2015 by John Wiley & Sons, Ltd.